SpringerBriefs in Applied Sciences and Technology

Forensic and Medical Bioinformatics

W0090647

Series editors

Amit Kumar, Hyderabad, India
Allam Appa Rao, Hyderabad, India

More information about this series at http://www.springer.com/series/11910

Naresh Babu Muppalaneni
Vinit Kumar Gunjan

Editors

Computational Intelligence Techniques for Comparative Genomics

Dedicated to Prof. Allam Appa Rao
on the Occasion of His 65th Birthday

 Springer

Editors
Naresh Babu Muppalaneni
C.R. Rao Advanced Institute
  of Mathematics, Statistics
  and Computer Science
Hyderabad
India

Vinit Kumar Gunjan
Annamacharya Institute of Technology
  and Sciences
Kadapa
India

ISSN 2196-8845   ISSN 2196-8853   (electronic)
SpringerBriefs in Applied Sciences and Technology
ISBN 978-981-287-337-8   ISBN 978-981-287-338-5   (eBook)
DOI 10.1007/978-981-287-338-5

Library of Congress Control Number: 2014955995

Springer Singapore Heidelberg New York Dordrecht London

Printed on acid-free paper

Springer Science+Business Media Singapore Pte Ltd. is part of Springer Science+Business Media
(www.springer.com)

To Prof. Allam Appa Rao on the occasion of his 65th birthday

Foreword

This book is an enthusiastic contribution of the best research work in the field of transdisciplinary research and allied to the proceedings of International Conference on Computational Intelligence: Health and Disease (CIHD 2014) to be held at Visakhapatnam, India during December 27–28, 2014. The main objective of this conference is to create an environment for (1) cross-disseminating state-of-the-art knowledge to CI researchers, doctors and computational biologists; (2) creating a common substrate of knowledge that both CI people, doctors and computational biologists can understand; (3) stimulating the development of specialized CI techniques, keeping in mind the application to computational biology; (4) fostering new collaborations among scientists having similar or complementary backgrounds and sowed the seeds of transdisciplinary research for which Prof. Allam Appa Rao has spent more than 2 decades and brought these parts of India to the notice of transdisciplinarians of the world.

Yet another element is provided by many interesting historical data on diabetes and an abundance of colorful illustrations. On top of that, there are innumerable historical vignettes that interweave computer science and biology in a very appealing way.

Although the emphasis of this work is on diabetes and other diseases, it contains much that will be of interest to those outside this field and to students of Biotechnology, Bioinformatics, Chemistry, and Computer Science—indeed to anyone with a fascination for the world of molecules. The authors have selected a good number of prominent molecules as the key subject of their essays. Although these represent only a small sample of the world of biologically-related molecules and their impact on our health, they amply illustrate the importance of this field of science to mankind and the way in which the field has evolved.

We think the contributors can be confident that there will be many grateful readers who will have gained a broader perspective of the disciplines of diabetes and their remedies as a result of their efforts.

Hyderabad, India

Naresh Babu Muppalaneni
Vinit Kumar Gunjan

Prof. Allam Appa Rao, Ph.D. (Engg), SDPS Fellow

Preface

This volume appears on the occasion of Prof. Allam Appa Rao's 65th birthday. Close scientific companions, who have worked with Allam for many years, contributed to this volume. Dr. Allam Appa Rao was the first ever to receive a Ph.D. from Andhra University in Computer Engineering, then part of Electrical Engineering, in the year 1984. During his more than four decades of professional experience, such as Head of the University Computer Centre, Head of the Department of Computer Science and Systems Engineering, Chairman Faculty of Engineering, Principal of College of Engineering, Vice Chancellor JNTU Kakinada, and Director CRRao AIMSCS, he shared his wisdom with fellow engineers and scientists across the globe through his innumerable research papers published in international journals and international conference proceedings.

41 scholars were awarded Ph.D. degrees under his guidance, 4 are in process of adjudication, and another 11 are in progress. "Best Academician Award" and "Best Researcher Award" were bestowed on him by Andhra University. Andhra Pradesh Government accorded him "Best Teacher" award. International Accreditation Council for Quality Education and Research (IACQER) bestowed on him Distinguished Fellow Award. Indian Science Congress Association (ISCA) conferred him with "Srinivas Ramanujan Birth Centenary Award" Gold medal for his significant and life-time contribution to the development of Science and Technology in the country specifically in the area of Computational Biology, Software Engineering, and Network Security. Dr. Allam is a strict disciplinarian. Government of India, NCC wing rightfully conferred on him the Honorary Colonel of the regiment position on him.

Prof. Allam's research contributions in Computer Engineering are vital, varied, and vast. He believes that CE is the "third pillar" that supports research, along with the pillars of theory and experimentation. He firmly believes that CE is an essential component to implement and strengthen any country's technological, economical, and social goals. According to him, Information Technology (IT) is a symbiosis between CE and other disciplines and mediates new interactions among the technologies, in multiple levels and scales.

Prof. Allam believes that the recent human genome breakthrough is the result of the work of multi-disciplinary teams where cooperation from computer experts was a fundamental ingredient. He stresses that the cells and cellular systems require viewing them as information processing systems, as evidenced by the fundamental similarity between molecular machines of the living cell and computational automata, and by the natural fit between computer process algebras and biological signaling and between computational logical circuits and regulatory systems in the cell.

Prof. Allam's current research work is influenced by his fascination for the mysteries that remain in the catalog of human genes. He believes that a systematic approach to the acquisition, management, and analysis of information available in human genes and their study can greatly enhance the quality and efficiency of medicines.

Dr. Allam proposed new computing methods, and showed the way to better treatments of disease and better understanding of healthy life. Prof. Allam has published more than 100 papers in peer-reviewed international journals.

Prof. Allam along with Dr. U.N. Das holds one US Patent, number: 8,153,392 Filing date: 28 Mar 2008 Issue date: 10 Apr 2012.

Prof. Allam is hailed by his innumerable student friends across the globe as a man of unfathomable knowledge and by his friends as a wonderful human being.

This volume contains a selection of the best contributions delivered at the International Conference on Computational Intelligence: Health and Disease (CIHD 2014) held at Visakhapatnam, India during December, 27–28, 2014. This conference is organized by Institute of Bioinformatics and Computational Biology (IBCB), Visakhapatnam, India jointly with Andhra University and JNTU Kakinada.

CIHD 2014 is aimed to bring together Computer Professionals, Doctors, Academicians, and researchers to share their experience and expertise in Computational Intelligence. The goal of the conference is to provide Computer Science Professionals, Engineers, Medical Doctors, Bioinformatics researchers, and other interdisciplinary researchers a common platform to explore research opportunities.

A rigorous peer-review selection process was applied to ultimately select the papers included in the program of the conference. This volume collects the best contributions presented at the conference.

The success of this conference is credited to the contribution of many people. In the first place, we would like to thank Prof. Allam Appa Rao, Director, C.R. Rao AIMSCS who motivated and guided us in making this conference a grand success. Our sincere thanks to Dr. Amit Kumar, Editor for Springer Briefs in Applied Sciences and Technology, who helped us in bringing this series. Moreover, special thanks are due to the Program Committee members and reviewers for their commitment to the task of providing high-quality reviews.

We would like to thank **Prof. B.M. Hegde** (Padma Bhushan Awardee, Cardiologist & Former Vice Chancellor, Manipal University) who delivered the keynote address. Last, but not least, we would like to thank the speakers **Grady Hanrahan** (California Lutheran University, USA), **Jayaram B.** (Coordinator, Supercomputing Facility for Bioinformatics and Computational Biology,

IIT Delhi), **Jeyakanthan J.** (Professor and Head, Structural Biology and Biocomputing Lab, Alagappa University), **Madhavi K. Ganapathiraju** (Pittsburgh University, USA), **Nita Parekh** (International Institute of Information Technology Hyderabad (IIIT-H), Hyderabad, India), **Pinnamaneni Bhanu Prasad** (Advisor, Kelenn Technology, France), **Rajasekaran E.** (Dhanalakshmi-Srinivasan Institute of Technology, Tiruchirappalli), **Sridhar G.R.** (Endocrine and Diabetes Centre, Krishnanagar Visakhapatnam, India).

December 2014   Naresh Babu Muppalaneni
                                                                               Vinit Kumar Gunjan

Prime Minister of India Dr. Manmohan Singh conferring the Shri Srinivasa Ramanujan Birth Centenary Gold Medal Award (a life-time achievement award for the contributions made in the field of Computational Biology, Network Security, and Software Engineering) to Dr. Allam Appa Rao at the 98th Indian Science Congress held in Chennai, India on January 3, 2011

Prof. Allam Appa Rao being conferred the rank of Honorary Colonel Commandant by NCC

Committees

International Conference on Computational Intelligence: Health and Disease (CIHD 2014)

27–28 December 2014
Visakhapatnam, India

The International Conference on Computational Intelligence: Health and Disease (CIHD 2014) held at Visakhapatnam, India during December 27–28, 2014 is organized by the Institute of Bioinformatics and Computational Biology (IBCB), Visakhapatnam, India jointly with Andhra University and JNTU Kakinada.

Conference Secretary

Dr. P. Sateesh, Associate Professor, MVGR College of Engineering

Organizing Committee

Dr. Ch. Divakar, Secretary, IBCB
Prof. P.V. Nageswara Rao, Head of the Department, Department of CSE, GITAM University
Prof. P.V. Lakshmi, Head of the Department, Department of IT, GITAM University
Prof. P. Krishna Subba Rao, Professor, Department of CSE, GVP College of Engineering (Autonomous)
Dr. G. Satyavani, Assistant Professor, IIIT Allahabad
Dr. Akula Siva Prasad, Lecturer, Dr. V.S. Krishna College

Shri. Kunjam Nageswara Rao, Assistant Professor, AU College of Engineering
Shri. D. Dharmayya, Associate Professor, Vignan Institute of Information Technology
Shri. T.M.N. Vamsi, Associate Professor, GVP PG College

Advisory Committee

Prof. P.S. Avadhani, Professor, Department of CS and SE, AU College of Engineering
Prof. P. Srinivasa Rao, Head of the Department, Department of CS and SE, AU College of Engineering
Dr. Raghu Korrapati, Professor, Walden University, USA
Prof. Ch. Satyanarayana, Professor, Department of CSE, JNTU Kakinada
Prof. C.P.V.N.J. Mohan Rao, Professor and Principal, Avanthi Institute of Engineering and Technology
Dr. Anirban Banerjee, Assistant Professor, IISER Kolkata
Dr. Raghunath Reddy Burri, Scientist, GVK Bio Hyderabad
Dr. L. Sumalatha, Professor and Head, Department of CSE, JNTU Kakinada
Dr. D. Suryanarayana, Principal, Vishnu Institute Technology, Bhimavaram
Dr. A. Yesu Babu, Professor and Head, Department of CSE, Sir C.R. Reddy College of Engineering, Eluru
Dr. T.K. Rama Krishna, Principal, Sri Sai Aditya Institute of Science and Technology

Finance Committee

Shri. B. Poorna Satyanarayana, Professor, Department of CSE, Chaitanya Engineering College
Dr. T. Uma Devi, GITAM University
Dr. R. Bramarambha, Associate Professor, Department of IT, GITAM University
Smt. P. Lakshmi Jagadamba, Associate Professor, GVP
Smt. Amita Kasyap, Women Scientist, C.R. Rao AIMSCS

Publication Committee

Dr. Amit Kumar, Publication Chair, Director, BDRC
Dr. Kudipudi Srinivas, Co-chair, Professor, V.R. Siddhartha Engineering College
Dr. G. Lavanya Devi, Assistant Professor, Department of CS and SE, AU College of Engineering

Dr. P. Sateesh, Associate Professor, MVGR College of Engineering
Dr. A. Chandra Sekhar, Principal, Sankethika Institute of Technology
Dr. K. Karthika Pavani, Professor, RVR and JC College of Engineering

Website Committee

Dr. N.G.K. Murthy, Professor of CSE, GVIT Bhimavaram
Dr. Suresh Babu Mudunuri, Professor of CSE, GVIT Bhimavaram
Shri. Y. Ramesh Kumar, Head of the Department, CSE, Avanthi Institute of Engineering and Technology

Financing Institutions

Department of Science and Technology, Government of India, New Delhi
Grandhi Varalakshmi Venkatarao Institute of Technology, Bhimavaram, AP, India
KKR and KSR Institute of Technology and Sciences, Guntur, AP, India

Contents

Diversified Insulin-Associated Beta-Behavioral and Endogenously Triggered Exposed Symptoms (DIABETES) Model of Diabetes in India

P. Raja Rajeswari, Chandrasekaran Subramaniam and Allam Appa Rao

Abstract The objective of the paper was to propose a diversified insulin-associated beta-behavioral and endogenously triggered exposed symptoms (DIABETES) model due to multiple factors so as to suggest medical remedies to improve the avoidance of diabetes disease. The various causes and their chances toward the most common diabetes effects including deaths in the nation are modeled as an diversified insulin association. The regional, social, biological and cultural aspects of the Indian community are considered to model the diabetes that of different categories. The performance and the correctness of the model are determined by considering the heterogeneity due to different factors that are specified as proposed IAB-ETES process algebra. The model-driven approach needs restricted operations on the variables to supplement any health care information system. The individual human responsibility and societal awareness along with the health regulation acts can minimize the vulnerability of the disease if the information technology for biological system complies with the enforcement acts of the developing nation.

Keywords Diversification · Association · Endogenous · Exposed · Cultural · Causality · Causes and outcomes

P. Raja Rajeswari (✉)
Sri Venkateswara Hindu College of Engineering, Machilipatnam, India
e-mail: rajilikhitha@gmail.com

C. Subramaniam
EASA Engineering College, Anna University, Coimbatore, India
e-mail: chandrasekaran_s@msn.com

A. Appa Rao
C.R.Rao Advanced Institute of Mathematics, Statistics and Computer Science,
University of Hyderabad Campus, Hyderabad, India
URL: http://www.allamapparao.org

© The Author(s) 2015
N.B. Muppalaneni and V.K. Gunjan (eds.), *Computational Intelligence Techniques for Comparative Genomics*, Forensic and Medical Bioinformatics,
DOI 10.1007/978-981-287-338-5_1

1 Introduction

Diabetes is an outcome among the heterogeneous group of diseases affecting different parts of the human body, with different predisposing factors, prevalence, and treatment effects. Over time, diabetes meaning high blood glucose damages nerves and blood vessels, leading to complications such as heart disease, stroke, kidney disease, blindness, dental disease, and amputations. Other complications of diabetes may include increased susceptibility to other diseases, loss of mobility with aging, depression, and pregnancy problems. The government is taking suitable measures and adapting strategies to create awareness among the people and treatments to the scientific society [1]. Diabetes is often more common in the wealthier parts of the population of low-income countries, but there is also evidence that in some middle-income countries, it is now more common in poorer sections of society [2]. However, focusing only on which socioeconomic group has the most diabetes obscures the fact that even in low-income countries, diabetes is already very common in the poorest sections of society—especially in urban areas, where one in six, or more, adults has diabetes.

The vast majority of this increase will occur in men and women aged 45–64 years living in developing countries. According to Wild et al. [3] the "top" three countries in terms of the number of T2D individuals with diabetes are India (31.7 million in 2000; 79.4 million in 2030), China (20.8 million in 2000; 42.3 million in 2030), and the US (17.7 million in 2000; 30.3 million in 2030). Clearly, T2D has become an epidemic in the twenty-first century. The different socio, economic, and individual culture brings major variations not only in various incidences but also in the pattern of Diabetes. The mortality rate of male and female people who are affected due to diabetes is mainly due to consumption of junk food, tobacco leaves, and high calorie sugar food [4]. The consumption of tobacco leaves habit typically occurs early in life through imitation of a family member or peers. The main reason for this deadly disease emanating from the medical communities is due to regional differences in the prevalence of risk factor in India as a biggest democratic nation with population more than 1,000 million. 347 million people worldwide have diabetes. In 2004, an estimated 3.4 million people died from consequences of high fasting blood sugar [5]. More than 80 % of diabetes deaths occur in low- and middle-income countries [6]. World Health Organization (WHO) projects that diabetes will be the 7th leading cause of death in 2030. Healthy diet, regular physical activity, maintaining a normal body weight, and avoiding tobacco use can prevent or delay the onset of Type 2 diabetes [7]. The statistical figures are consistent with the number of affected patients presenting for medical care with more advanced disease condition in India when compared with developed countries with overall survival rate is also reduced. Hence, an early detection would not only improve the cure rate, but it would also lower the cost and morbidity associated with treatment. The focus of the work is to formally model the diabetes through the various causes and exposures from the biological, social, and individual perspectives.

The organization of the paper is as follows: Sect. 2 brings out the various categories of diabetes identified so far by the medical and scientific communities with biological information.

Section 3 focuses on the causality model in which the social activities carried out throughout India in the name of family programs become some of the causes or exposures toward the outcome. Section 4 proposes insulin-associated beta-behavioral endogenously triggered exposed symptoms (IAB-ETES) a process algebra where the bio, socio, and individual activities are represented as operations for the causes as well as the remedies with necessary lemmas and notations. Section 5 concentrates on the diversified aggregation of the causality model and the satisfaction of the association properties of them with lot of observations and results through graph. Section 6 concludes the work by exposing the limitations of the proposed model based on the sample size and amount of finer details in the biological and social domain of interest.

2 N Category Diabetes

N category discusses the involvement of pancreas node in the affected biological area. Type-1 diabetes is caused by a lack of insulin due to the destruction of insulin-producing beta-cells in the pancreas. In Type 1 diabetes, an autoimmune disease—the body's immune system attacks and destroys the beta-cells. Normally, the immune system protects the body from infection by identifying and destroying bacteria, viruses, and other potentially harmful foreign substances. There are different categories of diabetes so far identified by the medical communities in which they generalized the diabetes as NX, where NX implies that nearby destruction of insulin-producing beta-cells in the pancreas cannot be assessed and at the same time, information not known for the category N0. If the white blood cells called T cells attack and destroy, beta-cells have not spread to any nearby cells it is of N1. If the N1 diabetes has spread to one beta-cell on the same side of the pancreas across is belonging to N2. This category is being further subdivided is divided into three subgroups: The N2a type diabetes has spread to next beta-cell on the same side as the primary diabetes. The N2b diabetes has spread to two or more beta-cells on the same side as the primary diabetes. Staging is a system that describes the extent of diabetes in the body. The biological details or data are needed to map the oral cancer effect due to various individual exposures or causes. The different categories of diabetes can be mapped with different reasons or causes from outside the medical field that are to be exposed. Other types of diabetes are caused by defects in specific genes, diseases of the pancreas, certain drugs or chemicals, infections, and other conditions. Over time, high blood glucose damages nerves and blood vessels leading to complications such as heart disease, stroke, kidney disease, blindness, dental disease, and amputations. The individual working attitude, personal behavior, age, and gender can also be brought into the proposed diversified association of endogenously triggered symptoms based model.

3 Diversified Causality Model

Diabetes mellitus affects people of every age, race, and background and is now a major modern cause of premature death in many countries around the world, with someone dying from diabetes Type 2 every 10 s worldwide [8]. The increased knowledge about the etiology, pathogenesis, and natural history of Type 2 diabetes has lead to primary prevention becoming a reality. Although an unequivocally accepted consensus regarding the early pathogenesis is still lacking, preventive measures can be based upon the best current available knowledge. The rapidly increasing number of patients with Type 2 diabetes, the severity of the disease, its multiple and severe complications, and the increasing socio-economic costs emphasize the importance of immediate preventive actions. The current situation for Type 2 diabetes can be compared with the epidemic of coronary heart disease during the 1960s and 1970s in many industrialized countries [9–11]. Various types of diabetes are: *Type 1 diabetes*: This condition involves destruction of more than 90 % of the insulin-producing cells of the pancreas, causing the pancreas to cease making insulin or to make very little [12]. *Type 2 diabetes*: While the pancreas continues to produce insulin, or even higher levels of insulin, the body develops a resistance to the insulin, causing scarcity of insulin for the body's needs and blood sugar levels remain permanently too high. It tends to run in families and around 15 % of people over 70 have diabetes Type 2 [12]. *Gestational diabetes* develops during pregnancy. Left undiagnosed, serious side effects can injure the mother and affect the unborn child. It also increases the chances of getting cardiovascular diseases after 15–20 years from 1.5 to 7.8 times! Recently, loss of beta-cell-specific traits has been proposed as an early cause of beta-cell failure in diabetes. However, the molecular mechanisms that underlie the loss of beta-cell features remain unclear. In diabetes, beta-cells exhibit an impaired capacity to compensate for increased insulin demand [13], a defect that has been ascribed to both inadequate cellular capacity to secrete insulin [14] and beta-cell death [15]. Among the earliest defects observed in T2D patients is a reduced ability of beta-cells to secrete insulin in response to elevated blood glucose levels [14]. This impairment in glucose-stimulated insulin secretion has been attributed to defects in glucose sensing [16], mitochondrial dysfunction [17], and oxidative stress [18]. Thus, mounting evidence suggests that defects in multiple cellular processes can compromise beta-cell function and could be a factor in T2D development. Furthermore, hyperglycemia has been shown to impair the expression of genes that are important for beta-cell identity [19]. In any observational studies or computational reasoning, the two groups of exposures and their association with the outcomes are to be clearly declared.

The social behavior of a group of people and the individual behavior of a patient can be related through an outsider entity. For example, the hereditary factor for the individual can have a direct impact on the social behavior as well as the bad habits of that particular individual. There are several key risk factors that increase the potential for people to suffer from diabetes, and while some of them are not under your control (such as age and genetics), others are (such as food intake and exercise). The risk factors for Type 2 diabetes include: There are several key risk factors that increase the potential for you to suffer from diabetes, and while some of them are not under your control (such as age and genetics), others are (such as food intake and exercise). The risk factors for Type 2 diabetes include: Obesity based on body mass index, a BMI over 29 increases your odds of diabetes to one in four [20]. This can indicate a family gene predisposing you to diabetes [20]. People of Hispanic, African American, Native American, Asian, or Pacific Islander descent are at almost double the risk of white Americans [20, 21]. Up to 40 % of women who experienced gestational diabetes are at risk of developing diabetes Type 2 later in life [21]. Diet high in sugar, [22] cholesterol, and processed food. Irregular or no exercise.

Diversified is but one of many problems that plague studies of cause and effect. A covariate must or may have an independent association with both the domains. The three domains like biology, sociology, and individual human activity are converged to identify the dominant diversity. The causes and effects are listed in which diversity depends on the output parameter and the target population of inferences. Since Type 2 diabetes is a heterogeneous and multifactorial disorder, preventive measures must be based upon modification of several risk factors simultaneously. Otherwise, the potential for prevention remains incomplete and insufficient. The existing evidence, however, suggests that even a single intervention, e.g, increased physical activity in sedentary people or weight loss in the obese, can lead to a marked reduction in the risk of Type 2 diabetes [23, 24]. There are two components to the design of a prevention strategy: (i) A population-based strategy for altering the lifestyle and those environmental determinants which are the underlying causes of Type 2 diabetes in the entire population; and (ii) A high-risk strategy for screening individuals at high risk for Type 2 diabetes and bringing preventive measures to this group on an individual basis.

The absence of diversity, which is a condition in any population of exposures, is not sufficient to identify the sharp null hypothesis of no causal effects at the unit level. The measurement errors in the survey and the opinion mining processes may reveal some biases of comparable or even greater magnitude can arise when number of samples is increased to determine the exact association. The missing data, as well as from model-specified errors, may lead to unstable outcomes after careful identification of the parameters in the links between primary and supplementary relationships. The causal association of exposure with the outcome of interest using non-experimental or experimental data can be obtained through correct reasoning in the relational model of data recorded.

As per the previous studies for judging causal inference, the following three criteria simultaneously should be satisfied:

(i)   Exposure must be proceeded from outcome
(ii) A statistical association should be revealed between exposure and outcome that is any changes on exposure status yields changes on outcome, and
(iii) The apparent association must be valid. The three conditions should be met for a factor to be a diversity.

The other studies revealed that confounder factor should associate with exposure that is it should have imbalance distribution between the exposed and the non-exposed groups. The earlier work on confounding and reasoning on epistemological studies showed that the confounding variable should be an independent risk factor for outcome of interest and at the same time, there should be an inherent association present in both the exposed and the non-exposed groups making the association not be an intermediate pathway relation between exposure and outcome. The work is also to propose a outcome-driven process algebra to inculcate the different processes through a number of operations and their notations. The IAB-ETS process model tries to aggregate the variables with the help of covariant from other domain of cause without affecting the positive and negative association with other two domains of interest in the area of diabetes modeling.

4 IAB-ETES Model Process Algebra

The process model can be very well applied to the interaction of multivariables in different domain of interest (DOI) with timing perspective. Operations under consideration play a major role in evaluating the performance of the proposed process model. The factors that predict the occurrence and rapidity of decline in β-cell function are still largely unknown, but most studies have identified islet cell autoantibodies as predictors of future decline and age as a determinant of residual insulin production at diagnosis. Historical as well as recent clinical experience has emphasized the importance of residual insulin production for glycemic control and prevention of end organ complications. Understanding the modifiers and predictors of β-cell function would allow targeting immunological approaches to those individuals most likely to benefit from therapy. According to present knowledge, the known high risk (symptoms for getting diabetes) of individuals are: (i) those with a family history of Type 2 diabetes; (ii) women who had gestational diabetes; (iii) people whose blood glucose has been previously found to be moderately increased; and (iv) hypertensive subjects. In addition, obese and physically inactive people have an increased risk of Type 2 diabetes. Altogether, these high-risk individuals are so numerous in modern societies that they would in fact comprise a large proportion of the adult population worldwide. As knowledge of the genetic predisposition for Type 2 diabetes increases, communities with a high genetic predisposition should be targeted.

The earlier PEPA model has been fully applied to enhance the performance of the said processes. The operators are identified to match all the three different domains of interest with equal weightage. The risks and the likelihood of any event or process can be linked to that particular domain of interest. The exposures or causes in one DOI have to be composed in a parallel way or not in a parallel way. The timing of that activity is to be taken care since the pre- or post-execution of one operation with other may not yield the expected outcome. The important operations with their explanations are given, and the corollaries are given to supplement the need or the importance of the said operation to map the outcome. The information fusion of multiple sources of evidences or exposures can be composed for an inferential reasoning. The operators can be applied sequentially or parallel to the set of exposures or even to a single cause to determine the risk-free outcome.

The essential operations in the proposed process model are as follows: (i) compose, (ii) confound, (iii) collapse, (iv) precede, (v) succeed, (vi) parallel, and (vii) not parallel always.

(i) *Compose*: This refers to the composition of all the possible exposed events in the multiple domain of interest.

Corollary *An entity in the Diabetes information domain is possible by composing an activity in bio domain with an event in social domain when mapped with corresponding entity in the information domain.*

It can be composed as a combination of maximum all the three events or any two events from two different domains.

$$\text{Probability}(\text{Diabetes}) = \sum_i \sum_m \sum_n B_{i.} \cdot S_m \cdot I_n$$

or

$$\sum_m \sum_n S_m \cdot I_n$$

where the subscripts i, m, and n are representing the variables in bio, socio and individual domains.

Lemma 1 *The Diabetes may be due to the occurrences of any biological event such as age or/and any social event such as consuming high calorie junk food, paan, or tobacco regularly which is being detected from the individual cause like bad habits within the inferential system.*

(ii) *Confound*: Generally speaking, Z is a confounder of X and Y if and only if it influences both X and Y where X, Y, and Z are mutually exclusive domain of interest (DOI).

Corollary *It can be said that an entity in the set Sociology is a confounder of the other sets, Biology and Individual if and only if socio entity influences both others.*

Lemma 2 *One or many entities many have same association with other entities from different domains. The next operators of interest is collapsing the association that has been existing before between any two DOIs.*

(iii) *Collapse*: This criterion can be applied under which an association remains invariant to the omission of certain variables from the DOI, say bio and socio.

Corollary *The dot product of B_i, biological DOI and S_m, sociological DOI yields again the dot product of B_p and S_q*

$$B_i \cdot S_m \mapsto B_p \cdot S_q$$

where $p = (n - i)$, $q = (j - m)$ where n, m are the total number of variables in respective DOI and

$$B_i + S_m \mapsto B_{p'} + S_{q'}$$

That is the (.) association between any two entities may lead to association of some other entities in the same DOI and the (+) association between any two entities may lead to association of some other entities in the same DOI but not the same as the dot product variable p and q.

Lemma 3 *The association between two entities may be additive or multiplicative in the sense that the sequence of operations are not important. With the timing perspective, the integration of two different activities like update and composition can also be introduced in the process model. They are like precedence and succession operators used in high level programming languages.*

(iv) *Precede*:

Corollary *The value of social and biological events are first updated and then composed with the biological system. The sequence of operations bear the priority of individual entities illustrating the recent events and entities are to be considered.*

$$+ + S_m \cdot B_i$$
$$+ + B_I \cdot S_m$$

Lemma 4 *This refers that the person might have been affected by oral cancer in the past but there are very many chances of he getting affected by the same in future.*

(v) *Succeed*:

Corollary *The socio events and the biological events are composed together and then, the result is updated and these operations can be represented as*

$$S_m B_i + + B_i S_m + +$$

Lemma 5 *This refers that presently if the person is suffering from oral cancer but there is no assurance that he might not be affected by the same in future.*

(vi) *Parallel Only*:

Corollary *The occurrence of the events causing the exposure may occur in parallel without any restrictions and many outcomes may occur parallel that can be represented as*

$$[S_m||I_n]$$

Lemma 6 *The entities in the DOI can associate with other entities many times parallel but only parallel not with any other entities or appear as a singleton entity.*

(vii) *Parallel Not Always*: The negation of this operation is used to detect or cure the disease where the two entities from the DOIs shall not appear parallel in the diagnosis or remedial activities. This can be represented as,

$$\sim [S_m||B_j].$$

The proposed process model consists of above mentioned actions for the purpose of identification, analysis, and diagnosis of the diabetes symptoms exposed in different DOI. The various associations and aggregations of these entities can be mapped to other relations in other domains for simplifying the decision making through logical reasoning.

In Table 1, the causes of diabetes due to various processes such as social, biological and individual is discussed and the effects of the process is also discussed. The basic need and motivation of this research work stems from the devastating statistics observed by health organizations around the world. In particular, there is a global warning given to Asian countries such as India, a developing

Table 1 Cause and effect of each processes

Process	Cause	Effect
Social	Social event (at all time)	• Cigarette smoking
		• Tobacco consumption
		• Energy drinks
		• Sports drinks
		• Junk foods (added sugar, saturated fats, and trans fats)
Bio	Cause (sometime)	• Hereditary
		• Obesity
		• Metabolic syndrome
Bio	Fungal infections, Cytomegalovirus, Rotavirus, Enteroviruses, Coxsackie virus B (CVB), Rubella	• Food habits (choosing high-fat, high-sugar, high-salt, or low-fiber foods)
Individual	Symptoms/treatment (always)	Type 1 diabetes, Type 2 diabetes, Gestational diabetes

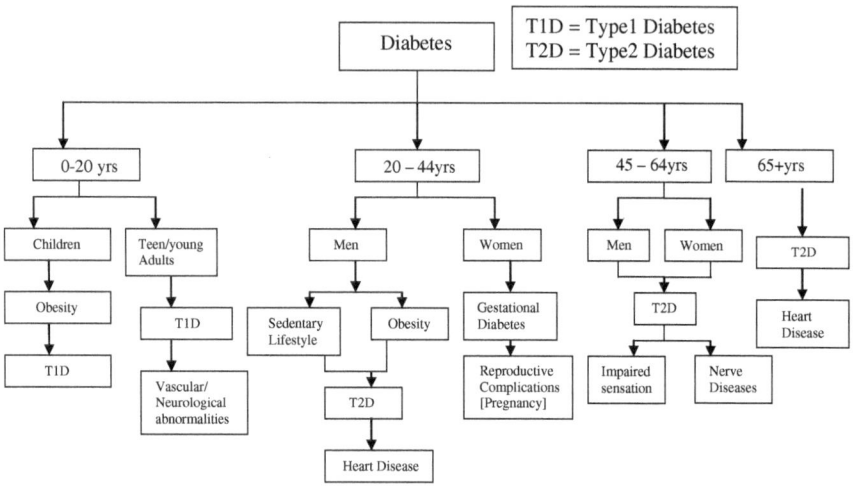

Fig. 1 Hierarchy of levels of diabetes

nation gets development not only in industrial setups but also in the diseased population. The vast majority of this increase will occur in men and women aged 45–64 years living in developing countries. The prevalence of T2D increases with age of population [3]. In developing countries, the largest number of people with diabetes are in the age group 45–64 years, while in developed countries, the largest number is found in those aged 65 years and over. These differences largely reflected differences in population age structure between developed and developing countries. Worldwide rates are similar in men and women, although they are slightly higher in men < 60 years of age and in women > age 65 years.

Of great concern is the recent increase in T2D in children [25]. A report based on the Pima Indians in Arizona noted that between 1967–1976 and 1987–1996, the prevalence of T2D increased 6-fold in adolescents [26].

In Fig. 1, it is shown how people at different age levels are affected by diabetes of which people in the range of 45–64 years are affected mostly.

Fig. 2 Mortality rate of diabetes

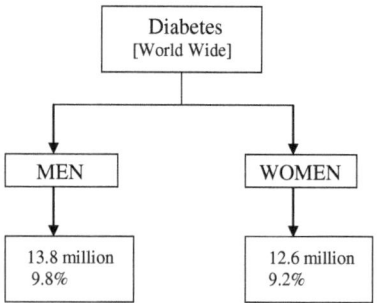

In Fig. 2, the mortality of diabetes rate based on worldwide statistics is divided, of which the male and female affected percentage is also shown. To plan and implement a massive health engineering strategy in the country, the foremost tasks are identification of the exact reasons for the deadly disease as per the regional and sentimental approaches. Different people with diversified social behavior makes the detectability of the exposure reasoning in this domain of diabetes modeling difficult. The solution will be correct and implementable once multiple confounders are identified across the entities.

5 Diversified Insulin Association Beta-Behavior

In the reasoning phase, it is necessary to understand the relationship between various entities of different DOI. In the object oriented style of modeling, the static structure of the various entities can be shown with a class diagram where the different instance of the structure can be connected. The class diagram is representing the direct and indirect aggregation of various classes into a main class. Any instance of one stack or pile of class may be connected to any other instance of different class as shown in Fig. 3. The overall class diagram representation is a multiple connections between the attributes from the three subclasses to the parent

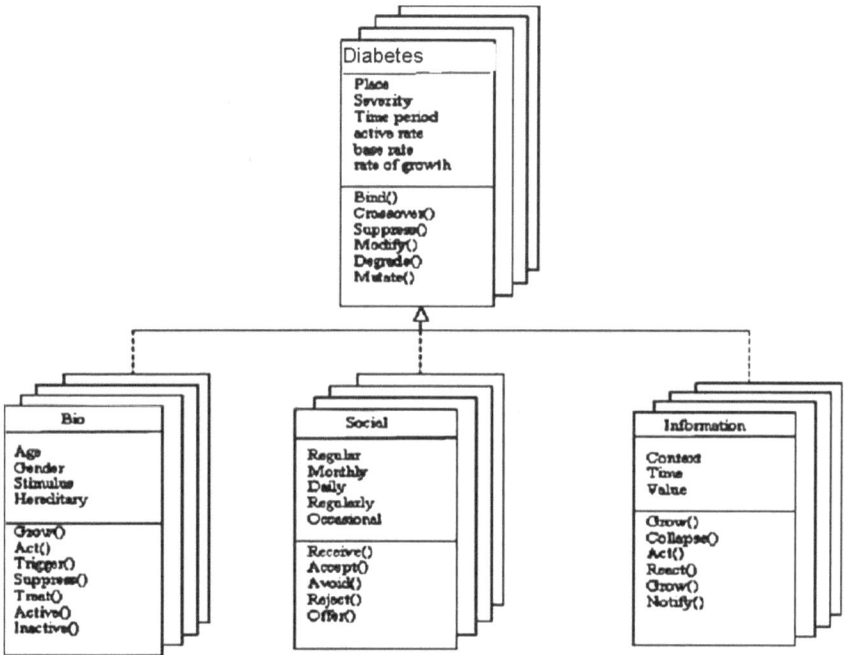

Fig. 3 Class diagram of diversified association

class. The pile of classes involved are diabetes stack, bio-pile, social, and individual pile of classes where each instantiation creates new association across the various member functions and attributes. The pile of classes helps to identify the exact or dominant confounder within the problem domain. To substantiate the aggregation of confounder, the numerical analysis is used with sample of bio, socio, and individual domain of interest. The following observations are made to identify the confounder and its collapsibility with respect to a set of entities shown in the form of Tables 2, 3 and 4.

From the Table 2, the bad habit which acts as a confounder for the bio-cause (hereditary) and social cause (Junk food consumption) has an impact value known as confounder impact (CI).

$CI_{\text{Bad Habit}} = 36,000/158,200 = 0.00275$

the confounder is having a positive impact with the positive confounder impact value,

$$CI_{\text{Bad Habit}} = 3200/20500 = 0.15609.$$

Table 4 shows that the confounder is having a negative impact with the negative confounder impact value, $CI_{\text{Bad Habit}} = 4,000/16,600 = 0.2409$.

Figure 4 shows the overall design diagram of the diversified insulin-associated beta-behavioral endogenously triggered exposed symptoms (DIABETES) model

Table 2 Association of hereditary and junk food

Hereditary	Junk food	
	Consume	No consume
Has	36,000	29,000
Has not	51,000	42,200
Total	87,000	71,200

Table 3 Association of hereditary and junk food with bad habit

Hereditary	Junk food	
	Consume	No consume
Has	3,200	4,300
Has not	10,000	3,000
Total	13,200	7,300

Table 4 Association of hereditary and junk food without bad habits

Hereditary	Junk food	
	Consume	No consume
Has	2,600	4,000
Has not	9,000	1,000
Total	11,600	5,000

Fig. 4 Diversified insulin-associated beta-behavioral endogenously triggered exposed symptoms model

where the social, biological and individual process features take various forms at different stages with respect to time and reason is shown. Initially, the bio-cause is identified by various symptoms social cause due to alert awareness and bad habits individual cause by bad habits and allergy which are aggregated together resulting in diabetes.

Table 5 represents the simulated values of the biological, social, and individual causes where the values are in the varying range.

Table 5 Simulated values for IAB-ETES causes

Time	Viral infections	Heart diseases	System aggregator	Affects kidneys	Diabetes
0.00	70.00	90.00	50.00	150.00	110.00
0.51	17.21	80.00	49.80	197.15	124.55
1.01	11.53	80.00	49.49	206.84	129.12
1.52	9.68	80.00	48.98	212.39	133.69
2.02	8.80	80.00	48.32	216.66	138.26
2.53	8.27	80.01	47.53	220.33	142.82
3.03	7.87	80.00	46.63	223.61	147.36
3.54	7.54	80.02	45.65	226.60	151.89
4.04	7.24	80.00	44.60	229.36	156.39
4.55	6.96	80.03	43.50	231.91	160.86
5.05	6.71	80.00	42.37	243.27	165.30

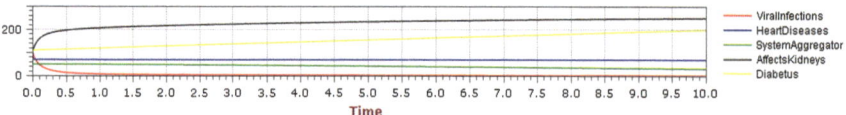

Fig. 5 Graphical representation for N-Diversity association

Figure 5 represents graphical data for the IEB-ETES causality model where the nodes for each cause are represented in graph by varying time period and their severity rates of it is shown diagrammatically.

6 Conclusion

The diversified insulin association with beta-behavioral in endogenously triggered exposed symptom model is proposed with different levels of association between different contributing factors. A bio-socioinformation approach is used to detect the probabilistic reasoning for diabetes in India. The BIO-SIN operations like confound, collapse, and parallel are also proposed to identify the contributing factor through different association and aggregation among the entities from three different domains of interest. The causality structure is achieved with a pile of stacked class diagram toward the deadly disease, and the reasoning is done computationally with the multiple insulin association. It is inferred from the proposed model, the diabetes may not be due to individual bad habits and will not be realized at the earliest time and it steadily shoots up. The many diversified causes initiate the symptom at the beginning and make the bio-system to accept only those habits at different ages of the two genders in India. The model concentrates on the endogenous-triggered behavioral bio-symptoms only. The model faces a limitations on the sample size of population surveyed and the time of observation. The fact that social and individual activities which also depend on the government policies and acts to govern or control those activities are not considered. These insulin-associated model has not included all variety of social behavior with correct quantification since India being the nation with a diversified culture. More detailed association model with fine tuning on different sections of people with their education will be undertaken as the future work.

References

1. Whiting D, Unwin N, Roglic G (2010) Diabetes: equity and social determinants. In: Blas E, Kurup A (eds) Equity, social determinants and public health programmes. World Health Organization, pp 77–94
2. Goldstein J, Jacoby E, del Aguila R et al (2005) Poverty is a predictor of non-communicable disease among adults in Peruvian cities. Prev Med 41(3–4):800–806
3. Wild S, Roglic G, Green A et al (2004) Global prevalence of diabetes. Diabetes Care 27:1047–1053

4. Shobhana R, Rao PR, Lavanya A et al (2000) Expenditure on health care incurred by diabetic subjects in a developing country–a study from southern India. Diabetes Res Clin Pract 48 (1):37–42
5. Hawkes C (2005) The role of foreign direct investment in the nutrition transition. Public Health Nutr 8(4):357–365
6. Knowler WC, Connor EB, Fowler SE et al (2002) Reduction in the incidence of type 2 diabetes with lifestyle intervention or metformin. The New England J Med 346(6):393–403
7. Lindström J, Louheranta A, Mannelin M et al (2003) The Finnish Diabetes prevention study (dps): lifestyle intervention and 3-year results on diet and physical activity. Diabetes Care 26 (12):3230–3236
8. Gillespie D (2008) Sweet poison: why sugar makes you fat. p 118, ISBN 978-0-670-07247-7
9. Keys A (1970) Coronary heart disease in seven countries. American Heart Association, New York
10. Uemura K, Pisa Z (1988) Trends in cardiovascular disease mortality in industrialized countries since 1950. World Health Stat Q 41:155–178
11. Sarti C, Rastenyte D, Cepaitis Z, Tuomilehto J (2000) International trends in mortality from stroke 1968 to 1994. Stroke 31:1588–1601
12. The Merck Manual of Medical Information (2003) Diabetes mellitus. p 962, ISBN 978-0-7434-7733-8
13. Cerasi E, Luft R (1967) The plasma insulin response to glucose infusion in healthy subjects and in diabetes mellitus. Acta Endocrinol (Copeh.) 55:278–304
14. Hosker JP, Rudenski AS, Burnett MA, Matthews DR, Turner RC (1989) Similar reduction of first- and second-phase B-cell responses at three different glucose levels in type II diabetes and the effect of gliclazide therapy. Metabolism 38:767–772
15. Butler AE, Janson J, Bonner-Weir S, Ritzel R, Rizza RA, Butler PC (2003) Beta-cell deficit and increased beta-cell apoptosis in humans with type 2 diabetes. Diabetes 52:102–110
16. Froguel P, Vaxillaire M, Sun F, Velho G, Zouali H, Butel MO, Lesage S, Vionnet N, Clément K, Fougerousse F et al (1992) Close linkage of glucokinase locus on chromosome 7p to early-onset non-insulin-dependent diabetes mellitus. Nature 356:162–164
17. Supale S, Li N, Brun T, Maechler P (2012) Mitochondrial dysfunction in pancreatic β cells. Trends Endocrinol Metab 23:477–487
18. Robertson RP (2004) Chronic oxidative stress as a central mechanism for glucose toxicity in pancreatic islet beta cells in diabetes. J Biol Chem 279:42351–42354
19. Jonas JC, Sharma A, Hasenkamp W, Ilkova H, Patanè G, Laybutt R, Bonner-Weir S, Weir GC (1999) Chronic hyperglycemia triggers loss of pancreatic beta cell differentiation in an animal model of diabetes. J Biol Chem 274:14112–14121
20. Fittante A (2007) Prevention's the sugar solution. p 264, ISBN 1-59486-693-7
21. CDC Diabetes and me. http://www.cdc.gov/diabetes/consumer/learn.htm
22. CDC Prevent diabetes. http://www.cdc.gov/diabetes/consumer/prevent.htm
23. Kuulasmaa K, Tunstall-Pedoe H, Dobson A et al (2000) Estimation of contribution of changes in classic risk factors to trends in coronary-event rates across the WHO MONICA Project populations. Lancet 355:675–687. (CrossRefMedlineWeb of Science http://bmb. oxfordjournals.org/external-ref?access_num=10703799&link_type=MED)
24. Eriksson KF, Lindgarde F (1991) Prevention of type 2 (non-insulin-dependent) diabetes mellitus by diet and physical exercise. The 6-year Malmö feasibility study. Diabetologia 34:891–898. (CrossRefMedlineWeb of Science http://bmb.oxfordjournals.org/external-ref? access_num=A1991GV96700006&link_type=ISI)
25. Bloom garden ZT (2004) Type 2 diabetes in the young: the evolving epidemic. Diabetes Care 27:998–1010
26. Fagot-Campagna A, Pettitt DJ, Engelgau MM et al (2000) Type 2 diabetes among North American children and adolescents: an epidemiologic review and a public health perspective. J Pediatr 136:664–672

Automatic Teaching–Learning-Based Optimization: A Novel Clustering Method for Gene Functional Enrichments

Ramachandra Rao Kurada, K. Karteeka Pavan and Allam Appa Rao

Abstract Multi-objective optimization emerged as a significant research area in engineering studies because most of the real-world problems require optimization with a group of objectives. The most recently developed meta-heuristics called the teaching–learning-based optimization (TLBO) and its variant algorithms belongs to this category. This paper provokes the importance of hybrid methodology by illuminating this meta-heuristic over microarray datasets to attain functional enrichments of genes in the biological process. This paper persuades a novel automatic clustering algorithm (AutoTLBO) with a credible prospect by coalescing automatic assignment of k value in partitioned clustering algorithms and cluster validations into TLBO. The objectives of the algorithm were thoroughly tested over microarray datasets. The investigation results that endorse AutoTLBO were impeccable in obtaining optimal number of clusters, co-expressed cluster profiles, and gene patterns. The work was further extended by inputting the AutoTLBO algorithm outcomes into benchmarked bioinformatics tools to attain optimal gene functional enrichment scores. The concessions from these tools indicate excellent implications and significant results, justifying that the outcomes of AutoTLBO were incredible. Thus, both these rendezvous investigations give a lasting impression that AutoTLBO arises as an impending colonizer in this hybrid approach.

R.R. Kurada (✉)
Department of Computer Science and Engineering, Shri Vishnu Engineering
College for Women, Bhimavaram, India
e-mail: ramachandrarao.kurada@gmail.com

K. Karteeka Pavan
Department of Information Technology, RVR & JC College of Engineering,
Guntur, India
e-mail: kanadamkarteeka@gmail.com

A.A. Rao
CRRao AIMSCS, UoH, Hyderabad, India
e-mail: apparaoallam@gmail.com

© The Author(s) 2015
N.B. Muppalaneni and V.K. Gunjan (eds.), *Computational Intelligence
Techniques for Comparative Genomics*, Forensic and Medical Bioinformatics,
DOI 10.1007/978-981-287-338-5_2

17

Keywords Automatic clustering · Teaching–learning-based optimization · Gene functional enrichments · Cluster validity indices

1 Introduction

Evolutionary algorithms (EA) are generic meta-heuristic optimization algorithms that use techniques inspired by nature's evolutionary processes. EA maintains a whole set of solutions that are optimized at the same time instead of a one single solution. The inherent randomness of the emulated biological processes enables them to provide good approximate solutions nevertheless. The recently emerged nature-inspired multi-objective meta-heuristic optimization algorithms teaching–learning-based optimization (TLBO) [1, 2] and its variations Elitist TLBO [3, 4] belong to this category. Both these algorithms aim to find global solutions for real-world problem with less computational effort and high reliability. The principle idea behind TLBO is the simulation of teaching–learning process of a traditional classroom in to algorithmic representation with two phases called teaching and learning. Elitist TLBO was pioneered with a major modification to eliminate the duplicate solutions in learning phase.

Clustering is the subject of active research in several fields such as statistics, pattern recognition, machine learning, data mining, and bioinformatics. The purpose of clustering is to determine the intrinsic grouping in a set of unlabeled data, where the objects in each group are indistinguishable under some criterion of similarity. Clustering is used to partition a dataset into groups, so that the data elements within a cluster are more similar to each other than data elements in different clusters. Automatic clustering addresses the challenge of determination the appropriate number of clusters or partitions mechanically.

Most of the existing clustering techniques, based on EA, accept the number of classes (k) as an input instead of determining the same on the iteration. Nevertheless, in many practical situations, the appropriate number of groups in a previously unhandled dataset may be unknown or impossible to determine even approximately. To avoid the algorithm struck in such blockage, automatic assignment of (k) value by the algorithm in each run was made tangible in this work. These automatic clusters are again endorsed with cluster validity indices (CVIs), which combine compactness and separability for assessing the quality of clusters. Cluster validity criteria are of three types external, internal, and relative. External indexes require a priori data for the purposes of evaluating the results of a clustering algorithm, whereas internal indexes do not. Internal indexes evaluate the results of a clustering algorithm using information that involves the vectors of the datasets themselves. The relative index evaluates the results by comparing the current cluster structures with other clustering schemes. The CVIs that are used in this work are rand index (RI) [5], advanced rand index (ARI) [5], Hubert index (HI) [6], silhouettes (SIL) [7], Davies and Bouldin (DB) [8], and Chou (CS) [9] measures, primarily finds the best partitioning in the underlying data.

This paper impersonate *k*-means clustering algorithm, procedures for automatic clustering, CVIs, visualization, and elitism techniques into TLBO. The objective of the novel AutoTLBO algorithm was to cluster the *Saccharomyces cerevisiae* categorized microarray datasets, and the expected multiple outcome was to attain optimal number of automatic clusters, mean values of CVIs, dendrograms and cluster profiles of co-expressed genes. These outcomes are assumed as summative assignment-I and is shown as Experiment 1 in Sect. 5. The obtained cluster profiles of AutoTLBO are used as inputs into Bioinformatics tools FatiGO for first opinion and database for annotation, visualization, and integrated discovery (DAVID) for second opinion. This verification procedure is named as summative assignment-II, primarily used to re-validate the results given by this novel AutoTLBO. Figure 1 unveils a broad road map of the proposed work in this paper. The input is fed into the tool in such a manner that the first list holds the gene-IDs of one of the cluster and the other list holds the gene-IDs of all the remaining clusters generated by this novel AutoTLBO algorithm. Two-stage preprocessing is imposed on the lists by applying statistical techniques such as Fisher exact test and duplicate elimination. Finally, these clean lists are used in gene ontology (GO) biological process to find the significant terms, term annotations % per list, *p*-value, FDRs, enrichment scores, etc. This entire set of test results of both the tools are publicized as Experiment 2 in Sect. 5. The re-validate techniques adopted in the tools manifest a positive sign that the novel AutoTLBO is power-packed in obtaining optimal number of automatic clusters and discrete gene cluster profiles. This silver lining absolutely ratify that the novel algorithm proposed in this work can used for attaining gene functional enrichments.

The rest of the paper is formed as follows. Section 2 exposes a basic background to the theme concepts used by other researchers, TLBO, and its variations.

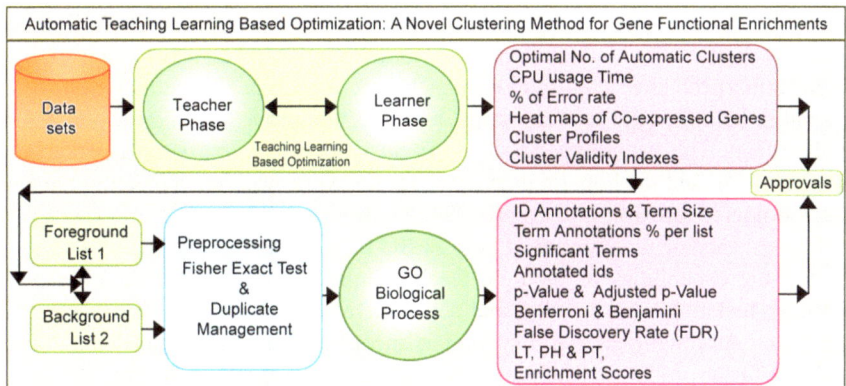

Fig. 1 Road map of AutoTLBO: a novel clustering method for gene functional enrichments

Section 3 summarizes the general framework of the TLBO algorithm. Section 4 presents the novel algorithm formulation. Results of a comparative study are presented in Sect. 5. Finally, conclusions are provided in Sect. 6.

2 Literature Survey

One of the most recent advancements in nature-inspired population-based meta-heuristics was the TLBO algorithm. This algorithm was initially proposed by Rao et al. in 2011 [1]. The worthiness of TLBO was successfully proved when it was applied over constraint real-time engineering optimization problems. In 2012, the same set of authors Rao et al. [2] have once again proven the efficiency on continuous non-linear large-scale problems with respect to the criterion best solution, average solution, convergence rate, and computational effort. A way to extend the TLBO algorithm to solve specific engineering optimization problems was also shown to the novel researchers by Rao et al. Hence, the authors of this paper embedded the concept of automatic clustering into TLBO and proved its effectiveness in clustering microarray datasets.

The methodology of this algorithm rolls between two entities teacher and learner. The outcome of the class learners is prejudiced by the teacher. Also, the superiority of the teacher is assessed by the learner's concert in requisites of marks and rating. The other significant factor that persuades the teacher's quality was the learner's progress among themselves by communication in the class. The same orientation was carry forwarded in this work to automatically cluster the genes stored in the form of microarray datasets.

The concept of elitist was introduced in to TLBO, and this modified version was released by Rao and Patel in 2012 [3] to solve constrained problems. The objective of this embedment was to eliminate the duplicate solutions and to get efficiency of algorithm when common variables such as population size and number of generations were used in the commencement of the algorithm. Again in 2013, Rao and Patel [4] pioneered elitism and common variable in TLBO to solve unconstrained engineering problems with different benchmark functions. The authors were successful in both [3, 4] to produce valuable results and prove the efficiency of TLBO with elitism. The inspiration for this present work was from the aforementioned articles [1–4], and the hopeful extensions to this work were implementing the new methodologies stated in articles [10, 11].

The enhancement incorporated in to TLBO by Rao and Patel in 2013 [10] was to exploit the capabilities of multiple teachers into classrooms (learners), adaptive teaching factor, tutorial training and self motivated learning. All these characteristics were thoroughly assessed in solving unconstrained multi-dimensional, linear, and non-linear problems. The most up-to-date work of Rao and Waghmare was in 2014 [11] which evaluated and produced efficient results by introducing multi-objective optimization with multiple trade-off in to TLBO over a set of the constraint and unconstrained functions.

TLBO relevance to cluster analysis was shown in 2012 by Amiri [12]; this study was accomplished by testing on quite a few numbers of datasets. Automatic clustering in multi-objective optimization framework using differential evolution was shown by Suresh et al. in 2011 [13]. The experimental result over different datasets proves the variations of DE that are desired for doing automatic clustering. Automatic clustering using genetic algorithms and generating optimality with Pareto front is well demonstrated by the same set of authors Suresh et al. in 2009 [14]. Cluster evaluation, ranking, and validation using CVIs are effectively shown by Liu et al. in 2005 [15]. The conceptualization toward fitting in automatic clustering into TLBO in the paper was from [13, 14].

Satpathy et al. in 2013 [16] brought an improved version of TLBO by using orthogonal design. This change was proved as a statistically effect method to generate an optimal offspring in EA. In the recent past, automatic clustering in TLBO was shown by Naik et al. in 2012 [17] using fuzzy c means. The results were well demonstrated over artificial and real datasets. In 2014, Murthy et al. [18] used automatic clustering in TLBO to find optimal number of clusters and shown potential results proving the efficiency of algorithm.

The proposal toward using automatic clustering in TLBO over microarray datasets was from the article published by Suresh et al. in 2009 [19] and Pavan et al. in 2011 [20]. Both these articles use a test suite to compare results over the gene datasets. The acquired optimal numbers of clusters are verified by using CVIs.

3 TLBO

TLBO algorithm is a teaching–learning methodology-motivated population-based algorithm, proposed by Rao et al. [1–4, 10, 11] which focused around the impact of a teacher on the after effect of learners in a class. In this optimization algorithm, the faction of learners are assumed as population and diverged configuration of variables are treated as distinctive subjects accessible to the learners, and their result is comparable to the fitness estimation value of this optimization issue. In the whole population, the best solution is treated as the teacher.

Teacher phase: It is included as the first segment of TLBO, where learners gain knowledge from the teacher. In this phase, the teacher attempts to increase the mean value of the class room from any value $mean_1$ to his or her echelon I_A. But sensibly it is not promising and a teacher can move the mean of the class room $mean_1$ to any other value $mean_2$ which is healthier than $mean_1$ depending on his or her competence. Considered $mean_j$ be the mean and I_i be the teacher at any iteration i. Now, teacher I_i will try to improve the existing mean $mean_j$ toward it so the new mean will be I_i designated as $mean_{new}$, and the difference between the existing mean and new mean is given as

$$\text{diverged\_mean}_i = r_i\left(\text{mean}_{\text{new}} - T_F * \text{mean}_j\right) \tag{1}$$

where T_F is the teaching factor that fixes the value of mean to be changed, and r_i is the random number in the range [0, 1], that is used to support the teaching factor. Value of T_F can either 1 or 2 which is an interrogative step, which is determined randomly with equivalent probability as:

$$T_F = \text{round}[1 + \text{rand}\,(0, 1)\{2 - 1\}] \tag{2}$$

The teaching factor is produced arbitrarily in TLBO within the scope of 1–2, in which 1 compares to no increase in the learning level and 2 relates to inclusive exchange of knowledge, and intermediate values indicate the exchange of knowledge. The shifting level of knowledge can be any depending on the learner competence.

Based on diverged_mean, the existing solution is updated according to the following expression:

$$\alpha_{\text{new},i} = \alpha_{\text{old},i} + \text{diverged\_mean}_i \tag{3}$$

Learner phase: It is included as the second segment of the algorithm, where learners improve their knowledge by communication among themselves. A learner adapts new things if the other learner has more knowledge than him. Precisely, the learning trend of this phase is articulated as follows:

At any iteration i, consider two distinct learners α_i and α_j where $i \neq j$.

$$\alpha_{\text{new},i} = \alpha_{\text{old},i} + r_i\left(\alpha_i - \alpha_j\right) \quad \text{if } f(\alpha_i) < f(\alpha_j) \tag{4}$$

$$\alpha_{\text{new},i} = \alpha_{\text{old},i} + r_i\left(\alpha_j - \alpha_i\right) \quad \text{if } f(\alpha_j) < f(\alpha_i) \tag{5}$$

3.1 Elitist TLBO Procedure

Step 1: *Initialization Stage*
         Initialize the population (learners), design variables (numbers of subjects offered to the learners) with random generation, threshold values, and termination criterion.
Step 2: *Elitist Teaching Phase*
         Select the best learners of each subject as a teacher for that subject and calculate mean result of learners in each subject.

   (a) Keep the elite solution
   (b) Calculate the mean of each design variable
   (c) Select the best solution
   (d) Calculate the diverged_mean and modify the solutions based on best solution

Step 3: *Elitist Teaching Phase—Update procedure amid with duplicate elimination*
Evaluate the difference between current mean result and best mean result according to Eq. (1) by utilizing the teaching factor T_F

(a) If the new solution is better than the existing solution, then accept or else keep the previous solution
(b) Select the solutions randomly and modify them by comparing with each other
(c) Modify duplicate solution via mutation on randomly selected dimensions of duplicate solutions before executing the next generation

Step 4: *Elitist Learners Phase*
Update the learner's knowledge with the help of teacher's knowledge according to Eq. (3)

(a) If the new solution is better than the existing solution, then accept or else keep the previous solution
(b) Replace worst solution with elite solution

Step 5: *Elitist Learners Phase—Update procedure amid with duplicate elimination*
Update the learner's knowledge by utilizing the knowledge of some other learners according to Eqs. (4) and (5).

(a) Modify duplicate solution via mutation on randomly selected dimensions of duplicate solutions before executing the next generation

Step 6: *Stoppage Criterion*
Repeat the procedure from Step 2 to Step 5 till the termination criterion is met.

(a) If termination criterion is fulfilled, then we get the final value of the solution or else repeat from Step 2 to Step 5.

4 Automatic Clustering Using Elitist TLBO (AutoTLBO)

The proposed automatic clustering using Elitist TLBO algorithm (AutoTLBO) follows a novel integrated approach by assimilation of elitism and cluster evaluation implanted into TLBO algorithm. Elitism is a mechanism used in this algorithm to preserve the best individuals from generation to generation. By this way, the algorithm never loses the best individuals found during the optimization process. In this algorithm, replacing the worst solutions with elite solutions is done at the end of learner phase. In the present work, duplicate solutions are modified by mutation on randomly selected dimensions of the duplicate solutions before executing the next generation. At the same time, the solutions are updated both in teacher phase and learner phase. The cluster evaluation procedure adopted in this algorithm is used to appraise the generated cluster with CVIs. The internal and external CVIs such as RI, ARI, HI, SIL, DB, and CS are used in this algorithm as an objective function to evaluate the cluster engendered.

4.1 AutoTLBO Algorithm

Step 1: *Initialization Phase*

Initialize each learner to contain Max k number of selected cluster centers and Max k (randomly chosen) activation thresholds in [0, 1]. Let α is a given dataset with n elements. The population α is initialized randomly. The dataset is generated with n rows and d columns using the following equation.

$$\alpha_{i,j}(0) = \alpha_j^{\min} + \text{rand}(1) * (\alpha_j^{\max} - \alpha_j^{\min}) \tag{6}$$

where $\alpha_{i,j}$ creates a population of learners or individuals. The ith learner of the population α at current generation t with d subjects is as follows:

$$\alpha_i(t) = \left[\alpha_{i,1}(t), \alpha_{i,2}(t), \ldots, \alpha_{i,d}(t)\right] \tag{7}$$

Step 2: *Teaching Phase*

Find the active cluster centers with value greater than 0.5, in each learner, and keep it as a elite solution as mentioned in Eq. 1

Step 3: *Teaching Phase—Update procedure amid with duplicate elimination*

For $t = 1$ to $t_{\max}$ do

(a) For each data vector α_p, calculate its distance from all active cluster centers using Euclidean distance or Euclidean metric

(b) Assign α_p to nearby cluster using simple k-means algorithm

(c) Modify duplicate solution via mutation on randomly selected dimensions of duplicate solutions before executing the next generation as described in Eqs. 2 and 3.

Step 4: *Learner Phase*

Evaluation of Clusters engendered with CVIs

(a) Evaluate each learner quality and find teacher, the best learner using RI or other indices

(b) Replace worst solution with elite solution

Step 5:   *Learners Phase—Update procedure amid with duplicate elimination*

    (a)   Update the learners according to the TLBO algorithm described in Eqs. 4 and 5

    (b)   Modify duplicate solution via mutation on randomly selected dimensions of duplicate solutions before executing the next generation

Step 6:   *Stoppage Criteria*

    (a)   Repeat the procedure from Step 2 to Step 5 till the termination criterion Step 6 is met.

    (b)   Report the final solution obtained by the globally best learner (one yielding the highest value of the fitness function) at time $t = t_{max}$.

In this experiment, Deb's heuristic constrained handling method [21] is used to handle the constraints. The rules are implemented at the end of the teacher phase (Step 3) and the learner phase (Step 5). Deb's method uses a tournament selection operator in which two solutions are selected and compared with each other. The following three heuristic rules are implemented on them for the selection:

- If one solution is feasible and the other is infeasible, then the feasible solution is preferred.
- If both the solutions are feasible, then the solution having the better objective function value is preferred.
- If both the solutions are infeasible, then the solution having the least constraint violation is preferred.

5 Experimental Analysis

The durability of AutoTLBO algorithm is assessed by compared various sized yeast datasets. The algorithm is implemented in MATLAB R2008a and run on a PC with Windows as OS and a core i3 processor operating at 2.93 GHz with 4 GB of RAM. All the results appeared in the analysis were the upshot of each dataset iterated with AutoTLBO algorithm after 50 independent runs. The following results correspond to the best solution attained by each algorithm, with respect to the coverage of assumed performance measures in the algorithm. The best results are displayed in boldface.

5.1 Experiment-1

The substantiated experimental analysis of the proposed algorithm is given in Table 1. Table 1 consolidates the results of *Saccharmoyces cerevisiae* categorized yeast datasets with distinct sizes and dimensions on AutoTLBO. The optimality and efficiency of the algorithm is estimated by 4 parameters, i.e., the core concept of

Table 1 Results of automatic clustering using Elitist TLBO in microarray datasets

Datasets	Size	Dim	No. of auto clusters	CPU time (s)	% of error rate	Mean value of cluster validity indices (CVIs)						
						ARI	RI	MIM	HIM	SIL	CSM	DB
Yeast238	238	19	4.2	25.07	5.801	0.921	0.972	0.928	0.944	0.951	1.646	1.069
Yeast384	384	19	5.4	40.84	8.937	0.952	0.907	0.893	0.914	0.829	1.696	1.088
Yeast2885	2885	19	6.4	1,700.75	38.411	0.821	0.810	0.890	0.721	0.891	2.912	1.505
Yeast2946	2946	18	5.6	1,100.80	35.477	0.851	0.879	0.821	0.859	0.726	2.325	1.502
Yeast4382	4382	25	6.2	3,510.24	44.278	0.721	0.663	0.737	0.826	0.747	3.702	2.016

engendering optimal automatic clusters, cluster profiles, minimum utilization of CPU time, low percentage of error rate, and mean values of CVIs marching toward its thresholds to justify automatic clustering accuracy.

In case of yeast238 dataset, the optimal number of automatic clusters is 4.2 which is close to assumed value, the percentage of error rate and CPU time is low. The mean value of CVIs SIL has a high optimality value and DB has a moderate value. This justifies that AutoTLBO can be applied for microarray datasets.

Figure 2 articulates the four cluster profiles with 39, 26, 43, and 129 respective gene-IDs in yeast238 dataset, and Fig. 3a fortitudes to a hierarchal representation of clustered expression profiles in yeast238 dataset as heat maps. This output exhales a result-oriented concrete outcome to impart new momentum and energy to Auto-TLBO for clustering microarray datasets.

The eligibility of the algorithm happens true when the dataset yeast384 is practiced over AutoTLBO. The algorithm attains a value of 5.4 as an optimal number of automatic clusters, with high favorable rate in ARI and DB indices' thresholds. An important observation was that the algorithm sustains the same minimal values both at error rate and CPU time. AutoTLBO is retrospective with belying expectations on yeast384 by leveraging five automatic clusters profiles with 68, 131, 45, 40, and 37 respective gene sizes. Figure 3b is fortitude in representing hierarchal cluster expression profiles on yeast384 dataset as a dendrogram. The result on yeast384 is a piecemeal but pragmatic with a promising growth that AutoTLBO is a landside in gene ontology with a corrective action.

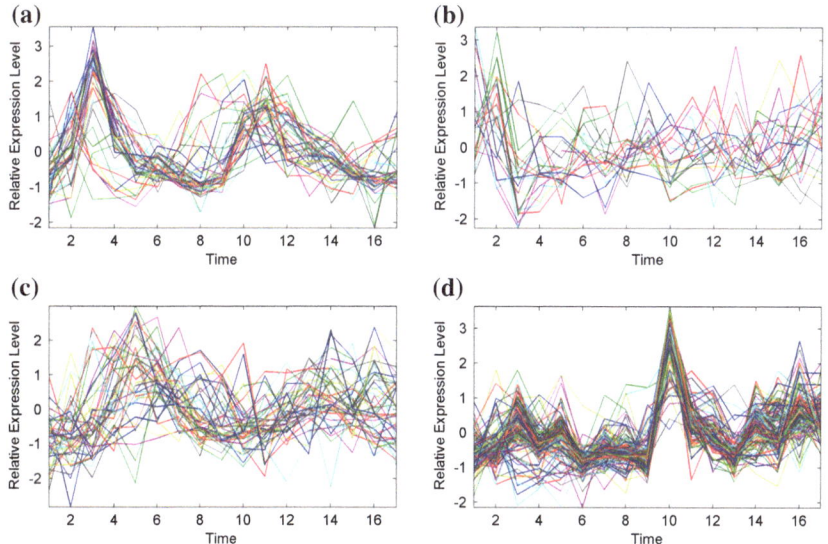

Fig. 2 Cluster profiles of co-expressed genes in yeast238 dataset. **a** cluster 4 of yeast238—cluster 1, **b** cluster 4 of yeast238—cluster 2, **c** cluster 4 of yeast238—cluster 3, **d** cluster 4 of yeast238—cluster 4

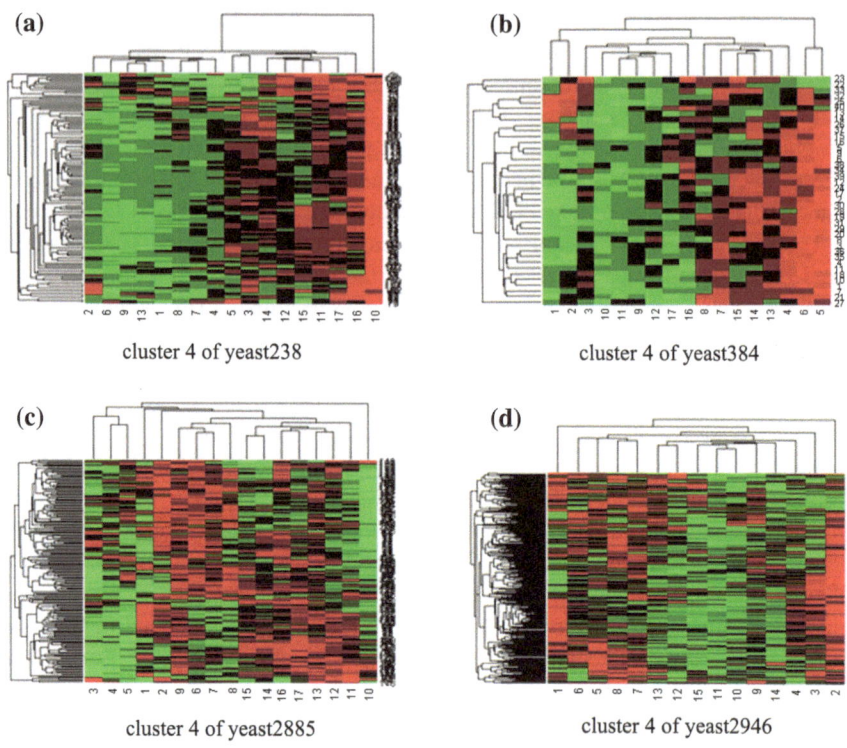

Fig. 3 Hierarchical expression profile of *S. cerevisae* categorized datasets. **a** cluster 4 of yeast238, **b** cluster 4 of yeast384, **c** cluster 4 of yeast2885, **d** cluster 4 of yeast2946

The quality of the algorithm is inspected by applying with a huge quantity dataset called yeast2885 with 2,885 gene-IDs and 19 dimensions. The efficiency of algorithm is exhibited by producing a value 5.4 as an optimal number of auto clusters, low usage of CPU time, and low percentage of error rate. Optimal threshold values are recorded at SIL and MIM indices to highlight the competence of clustering accuracy. Hence, this outcome launches AutoTLBO as a mere and potential immigrant in clustering microarray datasets. The core ideology of generating the automatic cluster when entrenched over yeast2885 is visualized as Fig. 4. This figure justifies that the five respective constellations of gene-IDs 1,422, 367, 877, 388, and 1,327 are prudent and consummated to the actual. Figure 3c is impinged with a heat map over the fourth cluster of yeast2885. The analysis from the heat map [22] was the data matrix holds the color information of microarray dataset along with numeric data. The red color is evidence for higher expression level of the gene, whereas green indicates low expression level and black indicates the absence of expression level. The perception was so bedazzle that the heat maps generated by AutoTLBO have higher expression level since most of the cluster is marked in red color.

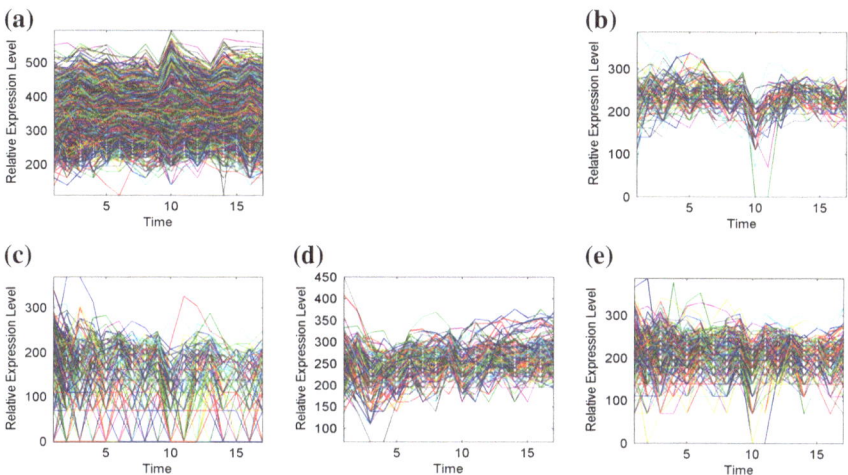

Fig. 4 Cluster profiles of co-expressed genes in yeast3 dataset with 2,885 instances. **a** dataset yeast2885—cluster 1, **b** dataset yeast2885—cluster 2, **c** dataset yeast2885—cluster 3, **d** dataset yeast2885—cluster 4, **e** dataset yeast2885—cluster 5

The investigation to estimate the capability of AutoTLBO is again protracted with almost on the same-sized dataset yeast2946, but with 18 dimensions. The proposed work makes its impact by producing the desired 5.6 automatic clusters with minimum percentage of error rate. The optimal mean CVI value is quoted in RI and HIM, proving the accuracy of automatic clusters despite the vast size of dataset. An important observation was yeast2885 consumes approximately 600 s of CPU time less when compared with yeast2946 dataset. The reason was the reduction in dimension of the microarray dataset makes that difference.

Figure 5 exhibits the inclusiveness of yeast2946 with six distinct sets of gene-ID of sizes 651, 193, 1,023, 656, and 422, respectively; also, the dispensation of fourth cluster expression profile is laid in as Fig. 3d. Both the figures exploit the importance of the proposed auto clustering method in the field of bioinformatics and recommends a revival strategy for the researchers to snoop into AutoTLBO.

The royalty of the work is exhibited when the algorithm is expended over the enormous volume of microarray dataset yeast5 with 4,382 instances of genes and 25 dimensions. Despite the size of the dataset the cluster accuracy sticks to the range of CVIs thresholds, by depicting an optimal value of 0.826 and 0.721, respectively, at HIM and ARI. AutoTLBO had a consistent eye watch over the percentage of error rate, but a moderate value of 44 % is recorded. The CPU time is coherent with the size of the dataset by producing accurate number of high volume clusters profiles. It is also noteworthy from AutoTLBO, to culminate five groups of discrete gene-IDs of 913, 967, 976, and 1,525 with respective sizes and obtain a tangible heat map. The hierarchical expression profile of the fourth cluster is visualized as shown in Fig. 6.

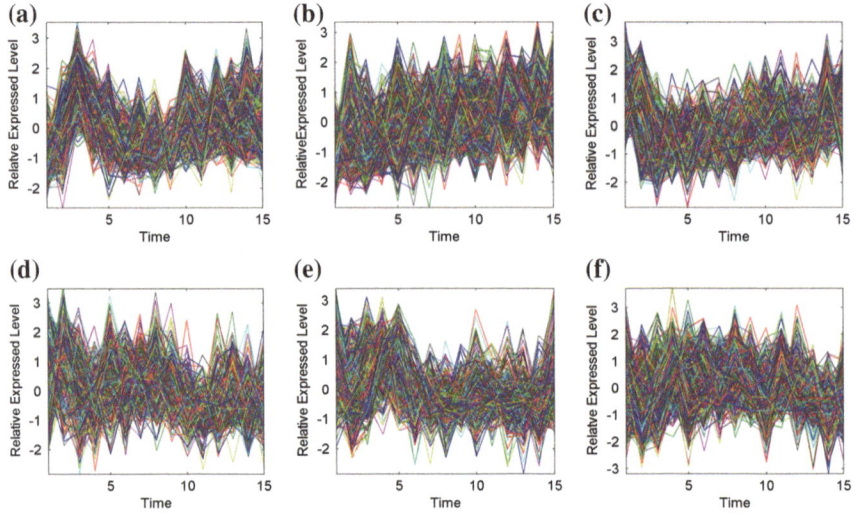

Fig. 5 Cluster profiles of co-expressed genes in yeast2946 dataset. **a** dataset yeast2946—cluster 1, **b** dataset yeast2946—cluster 2, **c** dataset yeast2946—cluster 3, **d** dataset yeast2946—cluster 4, **e** dataset yeast2946—cluster 5, **f** dataset yeast2946—cluster 6

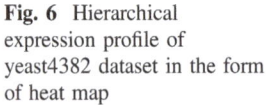

Fig. 6 Hierarchical expression profile of yeast4382 dataset in the form of heat map

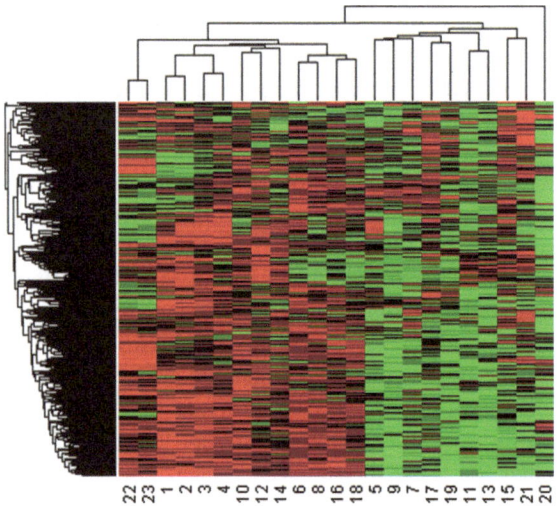

The upshots of the summative assessment-I are pragmatic by actioning all the constraints laid in Table 1. Hence, the promising results obtained by AutoTLBO over different-sized microarray datasets justify that the proposed work can be adopted in the field of bioinformatics to automatically cluster the gene profiles.

5.2 Experiment 2

This experiment focuses mainly to attain the gene functional enrichment of the incurred clusters of yeast datasets in Table 1 and re-validate the outcomes of AutoTLBO in experiment 1. The outputs of experiment 1 are used as inputs to the experiment 2. The Web-based functional annotation tools such as FatiGo [23] and DAVID [24] are used, particularly for gene-enrichment analysis. The clustered gene-IDs of each yeast datasets is segmented individually into two lists of genes, i.e., a group of interest as foreground and rest of genes as background lists. These lists are passed as inputs to the aforementioned tools for gene functional enrichment. The GO biological process is triggered on the gene lists to obtain significant results in range of selected level of gene ontology, gene annotations, p-value, etc. Since experiment 1 uses the yeast datasets, the genomics organism is treated as *S. cerevisiae*.

5.2.1 Outputs of FatiGo

Table 2 is inferred with the GO biological process applied between the levels 3–9 on all the five comparing datasets. The percentages of annotations produced by FatiGo in list 1 and list 2 of yeast234 and yeast384 datasets are outstanding. In the list 1 of yeast3 and yeast5, the percentage of annotations are 72 and 78.23 %, respectively, and in the list 2 of yeast3 and yeast5, the percentage of annotations are 73.48 and 77.66 %, respectively. This indicates that FatiGo has marked a reasonable good percentage of annotations on the list. The reason that was most looming was of the enormous size of data. The experimental results of Yeast2946 are not shown in this paper since the obtained results are merely equivalent to yeast2885. The clustering accuracy of TLBO was once again proven from column 6 and 7 of Table 2. The percentage of misclassification of the algorithm is zero in both in

Table 2 GO biological process on yeast datasets

GO biological process (levels from 3 to 9)						
Dataset	Total genes	ID annotations		Duplicate management		Number of significant terms
		List 1 annotations	List 2 annotations	List 1 duplicates	List 2 duplicates	
Yeast1	238	109 of 113 (96.4 %)	101 of 104 (97.12 %)	16 of 129 (0.12 %)	4 of 108 (0.04 %)	75
Yeast2	384	10 of 131 (83.97 %)	155 of 190 (81.58 %)	0 of 131 (0 %)	0 of 190 (0 %)	11
Yeast3	2,885	956 of 1,327 (72 %)	1,144 of 1558 (73.48 %)	0 of 1,327 (0 %)	0 of 1,558 (0 %)	6
Yeast5	4,382	1,193 of 1,525 (78.23 %)	2,218 of 2,856 (77.66 %)	0 of 1,525 (0 %)	0 of 2,856 (0 %)	4

Term	Term size	Term size (in genome)	Term annotation % per list		Annotated ids		Odds ratio (log e)	pvalue	Adjusted pvalue
translation (GO:0006412)	310	416	list 1:	4.75%	list 1: YAL016W, YCR003W		-0.5684	0.00005055	0.01377
			list 2:	8.09%	list 2: YBR04BW, YBR118W, YBR1				
ribosome biogenesis (GO:0042254)	294	379	list 1:	8.89%	list 1: YAL038C, YBL079W		0.4675	0.0002218	0.02278
			list 2:	5.76%	list 2: YBR04BW, YBR167C, YBR1				
RNA metabolic process (GO:0016070)	828	1139	list 1:	23.81%	list 1: YAL051W, YBL066C		0.4394	7.223e-8	0.00003936
			list 2:	16.76%	list 2: YAL029C, YAL0MPW, YBL0				
response to DNA damage stimulus (GO:0006974)	196	272	list 1:	6.33%	list 1: YAR007C, YBL003C		0.5739	0.0001692	0.02278
			list 2:	3.67%	list 2: YBL051C, YBL088C, YBR0				
alcohol metabolic process (GO:0006066)	163	217	list 1:	2.19%	list 1: YBR055C, YCR036W		-0.7198	0.0002507	0.02278
			list 2:	4.39%	list 2: YAR015W, YBR196C, YBR2				
RNA processing (GO:0006396)	328	462	list 1:	9.87%	list 1: YBR247C, YCL054W		0.4628	0.0001311	0.02278
			list 2:	6.45%	list 2: YAL029C, YBR167C, YBR2				

Fig. 7 Gene functional enrichment in yeast2 dataset with 2,885 genes

Term	Term size	Term size (in genome)	Term annotation % per list		Annotated ids		Odds ratio (log e)	pvalue	Adjusted pvalue
translation (GO:0006412)	310	416	list 1:	4.79%	list 1: YAL016W, YCR003W ...		-0.5878	0.00001046	0.00285
			list 2:	8.3%	list 2: YDL033C, YDL063C, YDL2				
RNA metabolic process (GO:0016070)	828	1139	list 1:	22.69%	list 1: YAL051W, YBL066C		0.3684	0.000003771	0.002055
			list 2:	16.88%	list 2: YBL008W, YBL052C, YBR0				
RNA modification (GO:0009451)	50	70	list 1:	1.97%	list 1: YCL054W, YDL014W		1.0457	0.0002771	0.03776
			list 2:	0.7%	list 2: YDL033C, YDL036C, YDL1				
folic acid and derivative metabolic process (GO:0006760)	8	14	list 1:	0.52%	list 1: YER183C, YOR267C		1.7976931349e+304	0.000213	0.03776
			list 2:	0%	list 2: no ids				

Fig. 8 Gene functional enrichment in yeast2 dataset with 4,800 genes

yeast2885 and yeast4382 and very low in yeast238 and yeast384. The number of significant terms quoted by FatiGo was very low, since the clusters submitted by the algorithm were very precise. Hence, all these evidences establish AutoTLBO as a top-drawer in associating biological phrases.

Figures 7 and 8 present the significant terms of yeast2885 and yeast4382 when p-value is less than 0.05. The nominal p-value forecasts the significance of the enrichment score for a single gene set.

5.2.2 Outputs of DAVID

The approach adopted in FatiGo tool is also practiced in this DAVID bioinformatics tools. The p-value generated on the yeast dataset via this tool is much similar to the outputs attained in FatiGo. This tool is used to judge the dataset with few more additional parameters. The results of yeast2885 dataset in DAVID tool are shown as Fig. 9. An optimal functional enrichment score of 0.93 is obtained for the annotation cluster. In the preprocessing stage, the Fisher exact statistical test is used to obtain the clean lists. Multiple comparison solutions between the list of foreground and background gene-ID is given as the false discovery rate. The significant value of 0.1 is acceptable for screening, and a list of independent, continuous dependent, normal gene-ID list is prepared. The Benjamini test type holds a cutoff of p-value to 0.05 and expect 0.05 genes to be significant by chance, and Bonferroni holds a cutoff of

Annotation Cluster 1	Enrichment Score: 0.93			Count	P_Value	Fold Change	Bonferroni	Benjamini	FDR	LT,PH,PT
GOTERM_CC_FAT	cytosolic ribosome	RT	i	3	3.5E-2	8.5E0	6.5E-1	6.5E-1	2.6E1	8,97,2209
GOTERM_BP_FAT	regulation of translation	RT	i	3	4.5E-2	7.6E0	9.9E-1	9.9E-1	4.0E1	9,105,2382
GOTERM_BP_FAT	posttranscriptional regulation of gene expression	RT	i	3	5.3E-2	7.0E0	1.0E0	9.3E-1	4.5E1	9,114,2382
GOTERM_BP_FAT	regulation of cellular protein metabolic process	RT	i	3	5.5E-2	6.8E0	1.0E0	8.5E-1	4.7E1	9,117,2382
KEGG_PATHWAY	Ribosome	RT	i	3	6.2E-2	5.5E0	3.2E-1	3.2E-1	2.6E1	5,87,792
GOTERM_CC_FAT	cytosolic part	RT	i	3	6.2E-2	6.2E0	8.5E-1	6.2E-1	4.2E1	8,133,2209
GOTERM_CC_FAT	ribosomal subunit	RT	i	3	7.1E-2	5.8E0	8.9E-1	5.2E-1	4.7E1	8,144,2209
SP_PIR_KEYWORDS	ribosome	RT	i	3	7.6E-2	6.0E0	9.5E-1	9.5E-1	5.1E1	13,115,2968
GOTERM_MF_FAT	structural constituent of ribosome	RT	i	3	1.0E-1	4.8E0	9.6E-1	9.6E-1	5.9E1	9,141,2046
SP_PIR_KEYWORDS	ribosomal protein	RT	i	3	1.1E-1	4.9E0	9.8E-1	7.5E-1	6.4E1	13,140,2968
GOTERM_CC_FAT	ribosome	RT	i	3	1.3E-1	4.1E0	9.8E-1	6.4E-1	6.9E1	8,201,2209
SP_PIR_KEYWORDS	protein biosynthesis	RT	i	3	1.4E-1	4.1E0	1.0E0	7.6E-1	7.5E1	13,167,2968
SP_PIR_KEYWORDS	ribonucleoprotein	RT	i	3	1.6E-1	3.9E0	1.0E0	7.2E-1	7.8E1	13,176,2968
GOTERM_CC_FAT	cytosol	RT	i	3	1.6E-1	3.6E0	9.9E-1	6.5E-1	7.7E1	8,230,2209
GOTERM_MF_FAT	structural molecule activity	RT	i	3	1.8E-1	3.4E0	1.0E0	8.6E-1	8.2E1	9,200,2046
GOTERM_CC_FAT	ribonucleoprotein complex	RT	i	3	2.6E-1	2.6E0	1.0E0	7.8E-1	9.3E1	8,315,2209
GOTERM_BP_FAT	translation	RT	i	3	3.6E-1	2.1E0	1.0E0	1.0E0	9.9E1	9,373,2382
GOTERM_CC_FAT	non-membrane-bounded organelle	RT	i	3	5.4E-1	1.5E0	1.0E0	9.6E-1	1.0E2	8,537,2209
GOTERM_CC_FAT	intracellular non-membrane-bounded organelle	RT	i	3	5.4E-1	1.5E0	1.0E0	9.6E-1	1.0E2	8,537,2209

Fig. 9 Gene functional enrichment in yeast2 dataset with 2,885 genes in DAVID Tool

Annotation Cluster 1	Enrichment Score: 1.29			Count	P_Value	Fold Change	Bonferroni	Benjamini	FDR	LT,PH,PT
GOTERM_BP_FAT	regulation of translation	RT	i	4	8.4E-3	8.1E0	6.4E-1	6.4E-1	9.2E0	11,100,2224
GOTERM_BP_FAT	posttranscriptional regulation of gene expression	RT	i	4	1.0E-2	7.5E0	7.2E-1	4.7E-1	1.1E1	11,108,2224
GOTERM_BP_FAT	regulation of cellular protein metabolic process	RT	i	4	1.1E-2	7.4E0	7.3E-1	3.6E-1	1.2E1	11,110,2224
GOTERM_CC_FAT	ribosome	RT	i	4	3.1E-2	4.8E0	6.1E-1	6.1E-1	2.3E1	9,191,2071
SP_PIR_KEYWORDS	protein biosynthesis	RT	i	4	4.3E-2	4.6E0	8.7E-1	8.7E-1	3.4E1	15,160,2782
GOTERM_CC_FAT	ribonucleoprotein complex	RT	i	4	9.9E-2	3.0E0	9.6E-1	4.6E-1	5.9E1	9,303,2071
GOTERM_BP_FAT	translation	RT	i	4	2.0E-1	2.3E0	1.0E0	1.0E0	9.2E1	11,350,2224
GOTERM_CC_FAT	non-membrane-bounded organelle	RT	i	4	3.2E-1	1.8E0	1.0E0	8.1E-1	9.6E1	9,517,2071
GOTERM_CC_FAT	intracellular non-membrane-bounded organelle	RT	i	4	3.2E-1	1.8E0	1.0E0	8.1E-1	9.6E1	9,517,2071
Annotation Cluster 2	**Enrichment Score: 0.92**			**Count**	**P_Value**	**Fold Change**	**Bonferroni**	**Benjamini**	**FDR**	**LT,PH,PT**
GOTERM_CC_FAT	cytosolic ribosome	RT	i	3	4.9E-2	7.3E0	7.8E-1	5.3E-1	3.5E1	9,95,2071
KEGG_PATHWAY	Ribosome	RT	i	3	6.7E-2	5.2E0	3.4E-1	3.4E-1	2.8E1	5,85,742
GOTERM_CC_FAT	cytosolic part	RT	i	3	8.1E-2	5.5E0	9.2E-1	5.7E-1	5.1E1	9,126,2071
GOTERM_CC_FAT	ribosomal subunit	RT	i	3	9.5E-2	5.0E0	9.5E-1	5.3E-1	5.7E1	9,138,2071
SP_PIR_KEYWORDS	ribosome	RT	i	3	1.0E-1	5.1E0	9.9E-1	9.2E-1	6.5E1	15,110,2782
SP_PIR_KEYWORDS	ribosomal protein	RT	i	3	1.4E-1	4.2E0	1.0E0	9.1E-1	7.7E1	15,134,2782
GOTERM_MF_FAT	structural constituent of ribosome	RT	i	3	1.5E-1	3.9E0	1.0E0	1.0E0	8.0E1	11,135,1918
GOTERM_CC_FAT	cytosol	RT	i	3	2.0E-1	3.2E0	1.0E0	6.7E-1	8.5E1	9,217,2071
SP_PIR_KEYWORDS	ribonucleoprotein	RT	i	3	2.1E-1	3.3E0	1.0E0	9.4E-1	8.9E1	15,170,2782
GOTERM_MF_FAT	structural molecule activity	RT	i	3	2.7E-1	2.7E0	1.0E0	1.0E0	9.5E1	11,196,1918

Fig. 10 Gene functional enrichment in yeast2 dataset with 4,382 genes

p-value equals to 0.05, to identify 5 % of the genes. The statistically significant values of these parameters are shown as column 8 and 9 in Fig. 9. The list total, population list, and population total are genes are shown as column 11 in Fig. 9.

Figure 10 displays the cluster obtained on yeast4382 with DAVID. The gene enriched in annotation term is scored as 1.29, which is an optimal value for the dataset of 4,382 gene-IDs. Figure visualizes the annotated clusters of yeast4382, whose p-value is <0.05. The impressive outputs of different standard statistics for multiple comparison corrections such as Benjamini, Bonferroni, LT, PH, and PT are given in column 8, 9, and 11 of Fig. 10.

The upshots of this summative assessment-II are indeed noteworthy. Realistic results are achieved by the usage of bioinformatics tools because the input quality was magnificent. Hence, this re-validation rationalizes the results produced by AutoTLBO is superlative.

6 Conclusion

This paper articulates a novel automatic clustering algorithm using TLBO for achieving gene functional enrichments. The results of AutoTLBO were again re-validated using benchmarked bioinformatics tools. Both the assessments substantiate that the AutoTLBO algorithm underscored in this paper has accurate creditability in yielding impending outputs. Thus, the hybrid approach of using this AutoTLBO algorithm of engineering studies in bioinformatics datasets is practical. As this paper is classified only to *S. cerevisiae* organism, the future envisaged scope of this work was to introspect AutoTLBO on the other assemblies of molecular functions in the gene ontology biological process.

References

1. Rao RV, Savsani VJ, Vakharia DP (2011) Teaching–learning-based optimization: a novel method for constrained mechanical design optimization problems. Computer Aided Des 43:303–315. doi:10.1016/j.cad.2010.12.015
2. Rao RV, Savsani VJ, Vakharia DP (2012) Teaching–learning-based optimization: an optimization method for continuous non-linear large scale problems. Inf Sci 183:1–15. doi:10.1016/j.ins.2011.08.006
3. Rao RV, Patel V (2013) Comparative performance of an elitist teaching-learning-based optimization algorithm for solving unconstrained optimization problems. Int J Ind Eng Comput 4: 29–50. doi:10.5267/j.ijiec.2012.09.001
4. Rao RV, Patel V (2012) An elitist teaching-learning-based optimization algorithm for solving complex constrained optimization problems. Int J Ind Eng Comput 3:535–560. doi:10.5267/j.ijiec.2012.03.007
5. Rand WM (1971) Objective criteria for the evaluation of clustering methods. J Am Stat Assoc 66(336):846–850
6. Hubert Lawrence, Schultz James (1976) Quadratic assignment as a general data analysis strategy. Br J Math Stat Psychol 29(2):190–241
7. Rousseeuw PJ (1987) Silhouettes: a graphical aid to the interpretation and validation of cluster analysis. J Comput Appl Math 20:53–65
8. Davies DL, Bouldin DW (1979) A cluster separation measure. Pattern Anal Mach Intell IEEE Trans On 2:224–227
9. Chou C-H, Su M-C, Lai Eugene (2004) A new cluster validity measure and its application to image compression. Pattern Anal Appl 7(2):205–220

10. Rao RV, Patel V (2013) An improved teaching-learning-based optimization algorithm for solving unconstrained optimization problems. Sci Iranica D 20(3):710–720. doi:10.1016/j.scient.2012.12.005
11. Rao RV, Waghmare GG (2014) A comparative study of a teaching–learning-based optimization algorithm on multi-objective unconstrained and constrained functions. J King Saud University—Comput Inf Sci 26: 332–346. doi:10.1016/j.jksuci.2013.12.004
12. Amiri Babak (2012) Application of teaching-learning-based optimization algorithm on cluster analysis. J Basic Appl Sci Res 2(11):11795–11802
13. Suresh K, Kundu D, Ghosh S, Das S, Abraham A (2009) Automatic clustering with multi-objective differential evolution algorithms. In: Evolutionary computation, 2009, IEEE Congress on CEC'09. IEEE, pp 2590–2597
14. Kundu D, Suresh K, Ghosh S, Das S, Abraham A, Badr Y (2009) Automatic clustering using a synergy of genetic algorithm and multi-objective differential evolution. In: Hybrid artificial intelligence systems. Springer, Berlin, pp 177–186
15. Liu Yimin, Özyer Tansel, Alhajj Reda, Barker Ken (2005) Integrating multi-objective genetic algorithm and validity analysis for locating and ranking alternative clustering. Informatica 29:33–40
16. Satapathy SC, Naik A, Parvathi K (2013) A teaching learning based optimization based on orthogonal design for solving global optimization problems. SpringerPlus 2:130
17. Naik A, Satapathy SC, Parvathi K (2012) Improvement of initial cluster center of c-means using teaching learning based optimization. Procedia Technol 6:428–435. doi:10.1016/j.protcy.2012.10.051
18. Murty MR et al (2014) Automatic clustering using teaching learning based optimization. Appl Math 5:1202–1211. doi:10.4236/am.2014.58111
19. Suresh Kaushik, Kundu Debarati, Ghosh Sayan, Das Swagatam, Abraham A, Han SY (2009) Multi-objective differential evolution for automatic clustering with application to micro-array data analysis. Sensors 9:3981–4004. doi:10.3390/s90503981
20. Pavan KK, Rao AA, Dattatreya Rao AV, Sridhar GR (2011) Robust seed selection algorithm for k-means type algorithms. Int J Comput Sci Inf Technol (IJCSIT) 3(5). doi:10.5121/ijcsit.2011.3513
21. Deb Kalyanmoy (2000) An efficient constraint handling method for genetic algorithms. Comput Methods Appl Mech Eng 186(2):311–338
22. Wilkinson L, Friendly M (2009) The history of the cluster heat map. The American Statistician 63(2)
23. Al-Shahrour F, Minguez P, Tárraga J, Medina I, Alloza E, Montaner D, Dopazo J (2007) FatiGO+: a functional profiling tool for genomic data. Integration of functional annotation, regulatory motifs and interaction data with microarray experiments. Nucleic Acids Research 35 (Web Server issue):W91–W96
24. Dennis G, Sherman BT, Hosack DA, Yang J, Baseler MW, Lane HC, Lempicki RA (2003) DAVID: database for annotation, visualization, and integrated discovery. Genome Biology 4 (5):P3

A Comparative Study of Methodologies of Protein Secondary Structure

M. Rithvik and G. Nageswara Rao

Abstract All living organisms are made up of cells and each cell in its turn consists of certain protein consequences which exercise an important role in catalyzing the chemical reactions. So, a study of a protein structure becomes a search lamp in the diagnosis of a disease. When the percent identity between two protein sequences falls below 33 %, it necessities to carry out the analysis of protein secondary structure. Of the several methodologies developed to analyze the protein secondary structure, two methods proved to be sound-dictionary of secondary structure of proteins (DSSP) and Garnier, Osguthrope and Robson (GOR), even though the prediction accuracy of GOR V is 73.5 % due to hazards in its implementation, GOR IV is generally used in spite of its accuracy being only to 64.4 %.

Keywords Protein · Residue · Motif · DSSP · GOR

1 Introduction

Every human body is constituted with a certain amount of cells. The functioning of the human body depends upon the functioning of cells. Every cell consists of certain protein sequences. Proteins play a vital role in catalyzing the chemical reactions in all the living organisms. Proteins are formed by the combination of several amino acids. The information flow from a DNA sequence to the protein structure is as follows.

The information flows from DNA to RNA first. From RNA, we acquire a protein sequence. This protein sequence helps us in predicting a protein structure. So, protein sequence plays a key role in predicting the structure of a protein. By knowing the structure, we can find the function of the protein.

M. Rithvik (✉) · G. Nageswara Rao
AITAM, Tekkali, India
e-mail: rithvikmadugula@gmail.com

G. Nageswara Rao
e-mail: gnraoaitam@gmail.com

© The Author(s) 2015
N.B. Muppalaneni and V.K. Gunjan (eds.), *Computational Intelligence Techniques for Comparative Genomics*, Forensic and Medical Bioinformatics, DOI 10.1007/978-981-287-338-5_3

1.1 Necessity to Predict the Protein Structure

Predicted structures can be used in structure-based drug design. It can help us understand the effects of mutations on structure or function. Structural knowledge brings understanding of function and mechanism of an action. It can help in prediction of function.

2 Background

Protein structure is classified into 4 categories [1]:

1. *Primary structure*: This structure is formed by the linear sequence of amino acids. A single change in the amino acid sequence of hemoglobin will result in a disease
2. *Secondary structure*: This structure is formed by the combination of hydrogen bonds between amino acids from two particular elements alpha helices and beta sheets
3. *Tertiary structure*: This is a more compact globular shape with carbon-rich amino acids that is inside far away from the surrounding water
4. *Quaternary structure*: This structure is formed by the combination of two or more polypeptide chains into one module with several subunits

In this paper, we are going to predict the secondary structure of a protein with the help of two algorithms 1. DSSP and 2. GOR.

We are also visualizing that structure with the help of tools.

This kind of prediction gives us an accuracy of 70 % when compared to other normal algorithms.

3 Methodology

The secondary structure of protein mainly constitutes of two elements: (i) alpha helix (ii) beta sheet. The regular patterns of the H-bonds are formed between neighboring amino acids, and the amino acids have similar φ and ψ bonds.

This secondary structure has three regular forms 1. alpha helix 2. beta sheet 3. loops. We give the input as the amino acid sequences and the results of output will be the predicted structure. A protein consists of about 32 % alpha (↑) helices, 21 % beta (↓) sheets, and 47 % loops or non-regular structure that is it may be a motif or a fold. In the above figure, the rotation angle around the N-cα is called ϕ and the

Fig. 1 Represents the alpha
helix

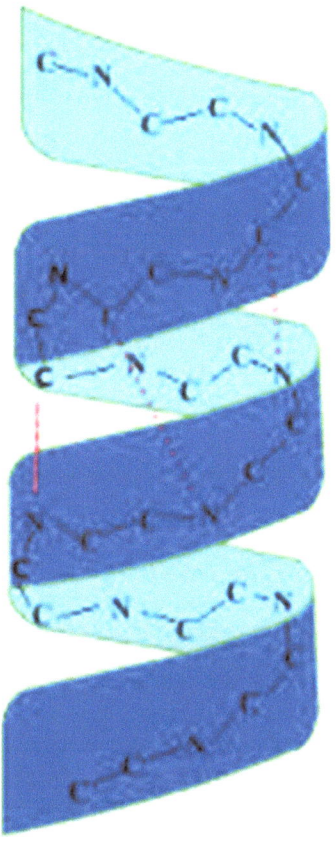

rotation angle between c-cα is called psy. The proteins that are evolved from a common ancestor are called homologous proteins. Generally, we used to predict the protein structure for a non-homologous protein from the protein data bank.

α Helix: we have to note an important point that here everything is related to residue and its position. The following are the various residues for a protein sequence [2]. It consists of 3.6 residues per turn and the angles between −60° and −50°, respectively. The hydrogen bond that exists between the NH group and CO group of residues is called α helices. It consists of a helical structure with a turn and an average of ten residues of length (Fig. 1).

β-sheets: β-sheet is built from a combination of several polypeptide chains that constitute of 5–10 residues long. These form a bond between CO and NH groups and form three kinds of chains: 1. parallel 2. antiparallel (Fig. 2).

In this paper, we are going to predict the protein structure by using the two methods 1. DSSP and 2. GOR.

(a) **(b)**

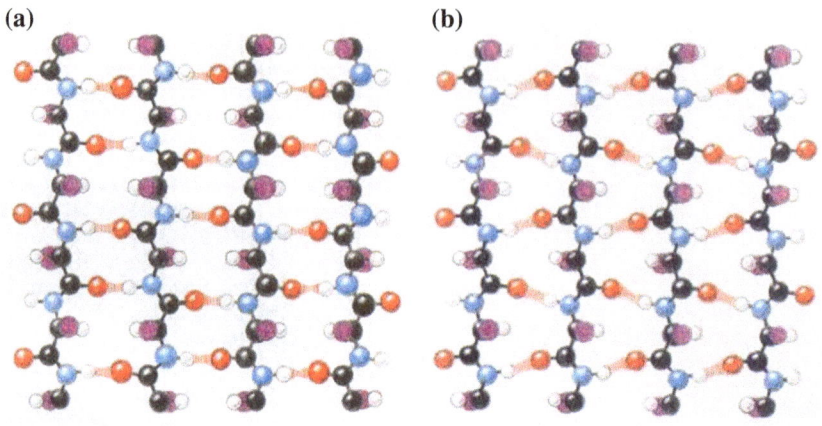

Fig. 2 a Represents the antiparallel beta sheet and **b** represents the parallel beta sheets

3.1 The DSSP

DSSP stands for dictionary of secondary structures. We generally take a non-homologous protein structure to calculate the protein structure. DSSP is a program that is used to convert or transfer the number of states that is it is used to decrease the number of states of residues from eight to three [3]. The reason behind the decrease of the number of states is to predict the secondary structure that consists of α helices, β-sheet, coil, and fold. It was first designed by Wolfgang Kabsch and Chris Sander to standardize the secondary structure assignment. DSSP is a database of secondary structure assignments for all protein entries in the protein data bank (PDB).

Brief history: The original DSSP application was written between 1983 and 1988 in the programming language PASCAL. This PASCAL code is automatically converted but maintaining this proved to be a lot of trouble-taking.

Working: The DSSP program works by calculating the most likely secondary structure assignment that was given in the 3D structure of a protein [2]. It does this by reading the position of the atoms in a protein followed by calculation of the H-bond energy between all atoms. The best two H-bonds for each atom are then used to determine the most likely class of secondary structure for each residue in the protein.

Structure: The structure of the DSSP is as follows:

G = 3-turn helix (3_{10} helix). Min length 3 residues.
H = 4-turn helix (α helix). Min length 4 residues.
I = 5-turn helix (π helix). Min length 5 residues.
T = hydrogen bonded turn (3, 4, or 5 turn)

E = extended strand in parallel and/or anti-parallel β-sheet conformation. Min length 2 residues.

B = residue in isolated β-bridge (single pair β-sheet hydrogen bond formation)

S = bend (the only non-hydrogen-bond-based assignment)

DSSP obtained from non-redundant PDB_select dataset, secondary structure assigned by DSSP into eight conformational states to three states H = HGI, E = EB, and C = STC.

GOR: The GOR method was first developed by Garnier, Osguthrope, and Robson [4]. It is an information theory-based method that uses a more powerful probabilistic technique of Bayesian inference. This method not only takes into account the probability of each amino acid having a particular secondary structure but also the conditional probability of the amino acid assuming each structure given the contributions of its neighbors. This method assumes that amino acids up to eight residues on each side that influence the secondary structure of the central residue. This program has many versions but here we are going to write the program in the fourth version. The algorithm makes use a sliding window of 17 amino acids. All possible pairs of amino acids in this window are checked for their information content as to predict the central amino acid by comparing the set of 266 other proteins of known structure. In this version, we are going to find the certain pair-wise combinations of amino acids that are present in the region called as flanking region. If a particular amino acid is surrounded by residues that prefer to be a helix, it is likely to be a helix, even if its individual helical preference is low. This method helps in considering the propensity of a single residue and calculates the position dependent propensities for helics, sheet and turn and thus simultaneously for all types of residues.

GOR V: This method is a bit improved method when compared to GOR IV method [5]. The improvement was made in the algorithm. It was inclusion of evolutionary information using PSI-BLASTMultiple alignments are generated using PSI-BLAST after five iterations based on the non-redundant database. The GOR V method prediction takes a longer time for the multiple alignments that is this kind of alignments takes much more time for hits.

Accuracy of the two methods: The prediction accuracy of the GOR IV method is 64.4 % and the prediction accuracy of GOR V method is 73.5 %. Even though GOR V method has more prediction among accuracy, people generally use the GOR IV method. This means that GOR V method is difficult to implement.

3.2 Results and Discussion

Here we are going to display the results of DSSP and GOR.

Output of DSSP

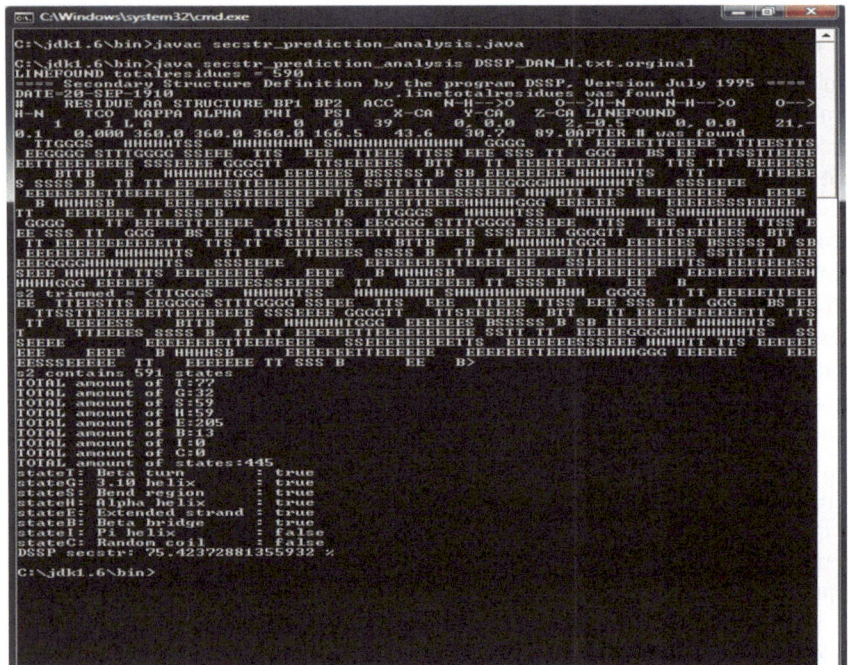

Output of GOR

See Fig. 3.

The results show in Fig. 3 represents the GOR IV output with number of α sheets, β sheets, loops and coils that helps in the prediction of secondary structure of a protein.

The protein structure can be visualized by using the various tools.

SWISS-PDB viewer: This is an application that provides a user-friendly interface allowing to analyze several proteins at the same time. This viewer helps us to visualize a protein secondary structure that was out when performing the DSSP and GOR programs.

Accuracy of prediction of protein structure: The accuracy of protein structure prediction depends upon the percentage of correctly predicted residues in sequences of known structure called Q3.

A method to calculate the accuracy is to calculate a correlation coefficient for each type of predicted secondary structure. The coefficient indicating the success of predicting residues in the α helical configuration cα is given by:

$$C\alpha = (p\alpha n\alpha - u\alpha o\alpha)/\text{square root of } [n\alpha + u\alpha][n\alpha + o\alpha][p\alpha + u\alpha][p\alpha + o\alpha]$$

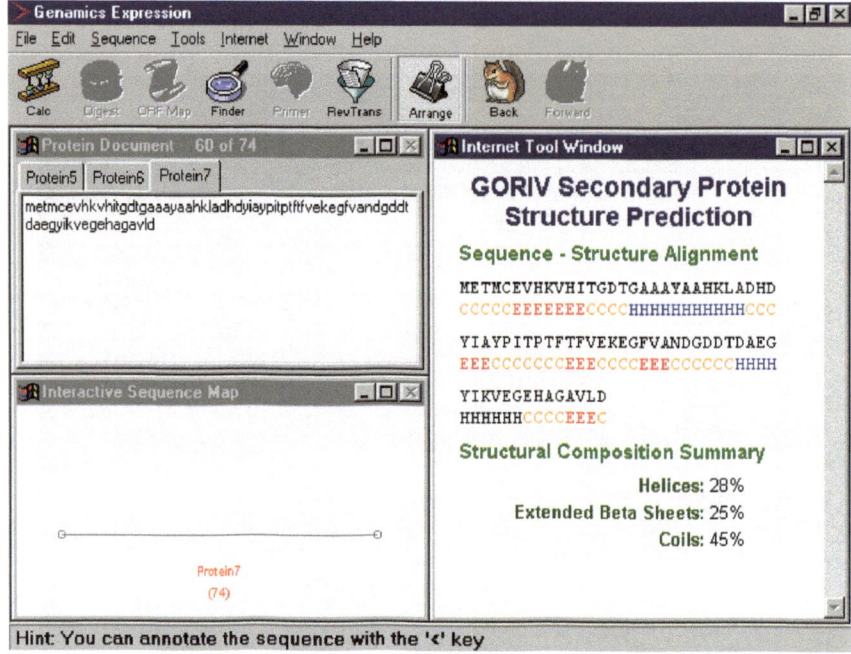

Fig. 3 Represents the GOR IV output

where

pα is the correct prediction
nα is the negative prediction
oα is the overpredicted positive prediction
uα is the number of unpredicted residues.

3.3 Conclusion

Here we are writing a program using DSSP and GOR methods to predict the protein secondary structure. Apart from the programs, we are going to visualize the protein secondary structure by using the SWISS-PDB viewer. The accuracy of DSSP algorithm is around 60 % and GOR IV is around 65 %, and when compared to the other programs, the GOR V gives around 80 % accuracy. Even though these two programs give us less accuracy, they are easy to write and implement in java. Even though new versions of the programs are giving more accurate results, people generally prefer these programs because these programs are easy in their implementation and robust in nature.

3.4 Future Scope

The further development of these programs helps us in predicting the more accurate protein secondary structures. The programs that will be developed in future must consider these programs in order to predict the secondary structure of the protein. The further applications of these programs must consider the Ramchandran plot in mind for their prediction accuracy.

References

1. Frishman D, Argos P (1995) Knowledge-based protein secondary structure assignment. Proteins 23(4):566–579. doi:10.1002/prot.340230412 (PMID 8749853)
2. Carter P, Andersen CA, Rost B (2003) DSSPcont: continuous secondary structure assignments for proteins. Nucleic Acids Res 31(13):3293–3295. doi:10.1093/nar/gkg626
3. Kabsch W, Sander C (1983) Dictionary of protein secondary structure: pattern recognition of hydrogen-bonded and geometrical features. Biopolymers 22(12):2577–2637. doi:10.1002/bip.360221211 (PMID 6667333)
4. Nageswara Rao G, Allam Appa Rao A java based protein secondary structure prediction program
5. Sen TZ, Jernigan RL, Garnier J (2005) GOR V server for protein secondary structure prediction. France received on Feb 23, 2005; revised on Mar 21, 2005; accepted on Mar 22 2005 advance access publication Mar 29 2005

A Sparse-modeled ROI for GLAM Construction in Image Classification Problems—A Case Study of Breast Cancer

K. Karteeka Pavan and Ch. Srinivasa Rao

Abstract Image segmentation is a process to determine regions of interest (ROI) in mammograms. Mammograms can be classified by extracting textural features of ROI using Gray Level Aura Matrices (GLAM). Scientists are selecting a fixed window size for all ROIs to find respective GLAM, though the masses will not occur in regular two-dimensional geometries. This paper makes an attempt to replicate the problem but by choosing arbitrary shape of masses as they occur. It is found that this kind of natural selection of the arbitrary shape yielded drastic reduction in time complexity by adopting the method suggested by us.

Keywords Classification · Mammogram · Texture · GLAM · Segmentation · ROI

1 Introduction

Breast cancer places a predominant role in women mortality [1]. Hence, this subject has become one of the hot topics in the study of image mining [2, 3]. Hither to scientists are choosing region of interest (ROI) as regular geometries while the fact is that the mass region will always be of arbitrary shape. For this reason, in their study, unconsciously, they are considering more number of unwanted pixels increasing time complexity in extracting features using gray level matrices [4, 5]. Therefore, in this paper, it is planned not to consider the unwanted pixels in feature extraction. The notion of sparse matrix is conveniently made used in extracting features based on the same technique of Gray Level Aura Matrices (GLAM) as

K.K. Pavan (✉) · Ch.S. Rao
R.V.R. and J.C. College of Engineering, Guntur, India
e-mail: karteeka@yahoo.com

Ch.S. Rao
e-mail: chereddy_sriny@yahoo.com

© The Author(s) 2015
N.B. Muppalaneni and V.K. Gunjan (eds.), *Computational Intelligence Techniques for Comparative Genomics*, Forensic and Medical Bioinformatics,
DOI 10.1007/978-981-287-338-5_4

adopted by the peer scientists [6, 7]. This novel idea has drastically reduced time complexity in constructing GLAM. The organization of the rest of the paper is as follows. Section 2 includes the motivation. Section 3 illustrates the methodology of sparse-modeled ROI. Sections 4 and 5 report experimental observations and conclusions.

2 Motivation

Many computer-aided classification systems are proposed for classification of masses of mammograms [8]. Identification of ROI using segmentation is the preliminary phase in the process of classification [9, 10]. Feature extraction and classification are the lateral steps of mammogram classification [11]. Textural feature extraction from ROI using gray level statistical matrices is the key issue in most of these systems [12]. The Gray Level Aura Matrix (GLAM) is a statistical method for extracting texture information from the images. The GLAM characterizes the spatial distribution of gray levels [13]. The size and shape of ROIs depend on the size and shape of suspected regions segmented in the preliminary phase. The size of ROI not only influences the computational time to construct the statistical matrix but also the classification rate. The scientists are choosing rectangular-shaped window for ROI. They are selecting different window sizes depending on their study and database that they have selected. For example, Chandy et al. [14] selected 200 × 200 pixels as uniform size of ROI. Mohanty et al. [15] fixed 50 × 50 pixels for ROI size for DDSM databases. In another study, Mohanty et al. [16] considered 256 × 256 pixels window for classification and detection of breast cancer. Hussain [17] used ROIs of size from 267 × 274 pixels to 1,197 × 1,301 pixels depending on size of the mass. For classification of mass, the authors have selected 1,024 × 1,024 pixels form mass center [18]. The regular rectangular shape and uniform size for all ROIs include more number of unwanted pixels which increases the computational time to construct gray level matrices, a feature extraction process. For example, in the Figs. 1, 2 and 3, no ROI is of rectangular shape. Total number of pixels of ROI of the first image is 2,191, second is 839 and third contains 68 pixels only. Therefore, we have proposed a novel idea not to consider the pixels outside the arbitrary shape of mass region. The model and algorithm explained in the next section.

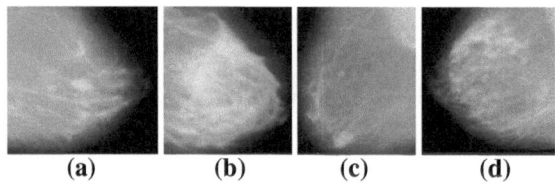

(a)   (b)   (c)   (d)

Fig. 1 Original mammograms

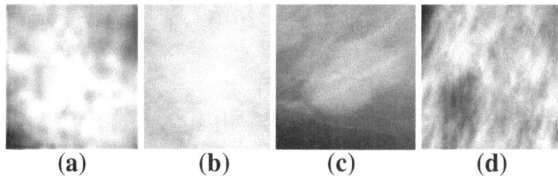

Fig. 2 ROIs of Fig. 1 (original mammograms)

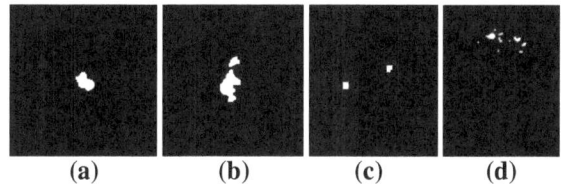

Fig. 3 Binary images of ROIs of Fig. 1 (original mammograms)

3 Methodology

3.1 Sparse-modeled ROI

The sparse model for arbitrary ROI is a sparse matrix with three columns: first column represents pixel value, and second and third are the corresponding x and y coordinates of the pixel in ROI. Example is shown in the Fig. 4.

Fig. 4 Sparse model for the arbitrary-shaped ROI

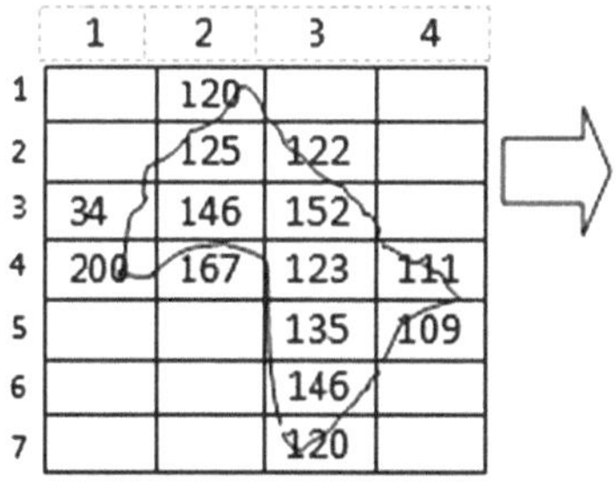

120	1	2
125	2	2
122	2	3
34	3	1
146	3	2
152	3	3
209	4	1
167	4	2
123	4	3
111	4	4
135	5	3
109	5	4
146	6	3
120	7	3

3.2 GLAM Algorithm for Sparse Model

An image can be modeled as a rectangular constitution S of $m \times n$ grids. Furthermore, consider an image s with a neighborhood system $N = \{N_s, s \in S\}$ can be defined. At which, the neighborhood N_s is built from the basic neighborhood E at pixel s. The basic neighborhood is thereby a chosen structural element [13]. Aura Measure: [13] Given two subsets A, B $\subseteq$ S, where $|A|$ is the total number of elements in A. The aura measure of A with respect to B for neighborhood system N is given as follows:

$$m(A, B, N) = \sum_{S \in a} |N_s \cap B|$$

GLAM: [13] let N be the neighborhood system over S and $\{S_i, 0 \leq i \leq G - 1\}$ be the gray level set of an image over S with G as the number of different gray levels, then the GLAM of the image $A(N)$ is as follows:

$$A(N) = [a_{i,j}] = [m(s_i, s_j, N)]$$

whereby $S_i = \{s \in S \mid x_s = i\}$ is the gray level set corresponding to the ith level, and $m(S_i, S_j, N)$ is the aura measure of S_i with respect to S_j with the neighborhood system N.

Because of the 16-bit resolution of the original image, the GLAM would be a matrix with a maximum size of $65,536 \times 65,536$. To reduce the size of the matrix and the necessary time for the retrieval, the ROI has to be quantized before the GLAM generation. As discussed in the results, the smallest possible number of allowed gray levels without loss of performance is eight. The result is a matrix with 64 entries which is transformed to a feature vector with 64 entries and normalized for the feature comparison. Because of the normalization, the GLAM gets independent from the size of the ROI. The details of the important variables in the pseudo code are as follows:

Input—g = sparse model for ROI,
Output—glam,
gl = minimum in pixels intensity values,
gh = maximum of pixels intensity values,
range = range of pixel values to be considered, i.e., gh−gl + 1.

Following is the MATLAB code for the construction of GLAM using sparse-modeled ROI

```
for k1=gl:gh
  for k2=gl:gh
    c=0;
    for i=1:mm
        if A(i,1)==k1
            if i+1<=mm && A(i+1,2)==A(i,2) && (i+1,3)==A(i,3)+1 && A(i+1,1)==k2
            c=c+1;
            end
            if i-1>=1 && A(i-1,2)==A(i,2) && A(i-1,3)==A(i,3)-1 && A(i-1,1)==k2
            c=c+1;
            end
            it=0;
            ind=i-1;
            while (ind>=1 && it<=colrange)
                if A(ind,2)==A(i,2)-1 &&    A(ind,3)==A(i,3)&& A(ind,1)==k2
                    c=c+1;
                    break;
                else
                    ind=ind-1;
                    it=it+1;
                end
            end
            it=0;
            ind=i+1;
            while (ind<=mm && it<=colrange)
                if A(ind,2)==A(i,2)+1 && A(ind,3)==A(i,3)&& A(ind,1)==k2
                    c=c+1;
                    break;
                else
                    ind=ind+1;
                    it=it+1;
                end
            end
        end
    end
    glam(k1-gl+1,k2-gl+1)=c;
  end end
```

4 Experimental Results and Discussions

We have conducted experiments on various mammograms available in Mammography Image Analysis Society (MIAS) database and also on the mammograms collected from the known radiologist. While in the experimentation, we have observed that the sparse-modeled ROI greatly reduces the time required to construct the GLAM than proper fixed window size. Here, we have fixed the rectangular or window size based on the minimum, maximum x, y coordinates of segmented mass

Table 1 Time complexity of 100 images

S. no	Time taken in seconds		S. no	Time taken in seconds		S. no	Time taken in seconds		S. no	Time taken in seconds	
	Fixed window	Sparse model		Fixed window	Sparse model		Fixed window	Sparse model		Fixed window	Sparse model
1	423.970638	0.044355	26	175.7225	65.25913	51	225.1796	104.3384	76	248.5533	154.8511
2	406.620913	0.2257724	27	163.0255	78.33145	52	146.8674	55.06624	77	282.0763	149.2762
3	314.827351	5.0535071	28	178.7782	91.13546	53	195.2088	90.43119	78	253.4304	151.2813
4	387.48534	34.731584	29	178.2487	99.83632	54	277.265	154.1731	79	212.4593	112.1969
5	305.560462	357.4795	30	139.9556	63.27879	55	264.3411	168.1003	80	256.0509	130.4261
6	243.899644	0.3699943	31	108.1272	34.39194	56	181.3133	89.30317	81	248.1469	148.1799
7	315.155052	385.79367	32	242.8082	142.5359	57	200.8823	112.1695	82	271.2852	141.797
8	295.155539	0.0083386	33	261.5322	169.9708	58	338.6756	252.855	83	256.0291	158.2177
9	395.824661	65.911515	34	166.4608	70.26944	59	262.5568	169.7577	84	244.628	158.9861
10	235.828732	187.46733	35	192.74	89.64735	60	408.2481	0.90761	85	327.5292	206.0257
11	154.351078	0.0378449	36	246.5741	143.6344	61	352.7393	254.4198	86	250.5986	164.0699
12	157.406497	154.04349	37	293.1976	178.048	62	221.8314	142.4193	87	419.6116	0.002353
13	457.720359	14.922187	38	188.3866	122.705	63	194.8638	102.3808	88	487.5725	425.4757
14	403.093725	19.708962	39	231.8158	136.2482	64	204.0718	100.7403	89	210.7851	124.4699
15	286.616527	1.4203566	40	493.6045	223.4936	65	178.6488	79.57717	90	260.0046	169.1708
16	258.284921	97.33222	41	480.4541	128.3134	66	236.2415	146.961	91	440.8307	72.64158
17	341.416155	5.5772605	42	330.332	220.1545	67	225.8924	144.8546	92	269.7271	54.15686
18	337.634549	26.373103	43	219.6698	136.1643	68	388.803	376.6609	93	449.4328	41.48826
19	387.618735	219.2832	44	149.7672	56.25259	69	301.762	217.7642	94	400.5187	89.68809
20	323.007328	86.619612	45	138.38	62.19115	70	230.5811	138.0605	95	448.565	24.88563

(continued)

Table 1 (continued)

S. no	Time taken in seconds		S. no	Time taken in seconds		S. no	Time taken in seconds		S. no	Time taken in seconds	
	Fixed window	Sparse model		Fixed window	Sparse model		Fixed window	Sparse model		Fixed window	Sparse model
21	273.020383	33.736677	46	240.1123	134.5153	71	229.0278	132.2187	96	260.482	52.12349
22	345.445736	148.76372	47	189.7236	109.0678	72	392.6921	135.3202	97	268.2461	82.81837
23	339.468135	25.636506	48	370.4142	0.398458	73	337.1127	61.48409	98	298.6097	20.9104
24	143.394236	63.892328	49	319.5939	225.3872	74	438.0198	46.80678	99	302.0604	113.9713
25	156.936412	71.986574	50	216.1657	110.5925	75	312.8642	200.7929	100	367.4819	68.20154

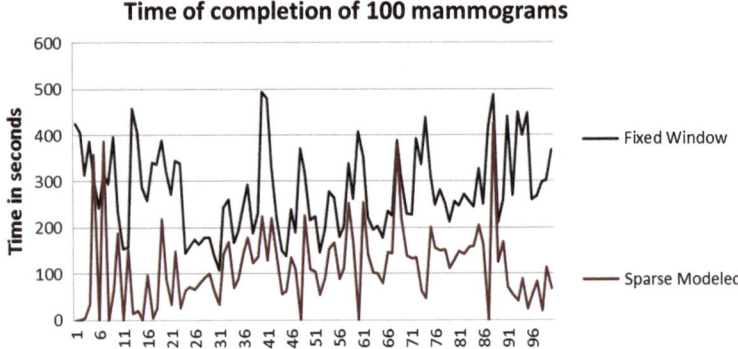

Fig. 5 Observed timings of 100 images

Fig. 6 Average times of two
models

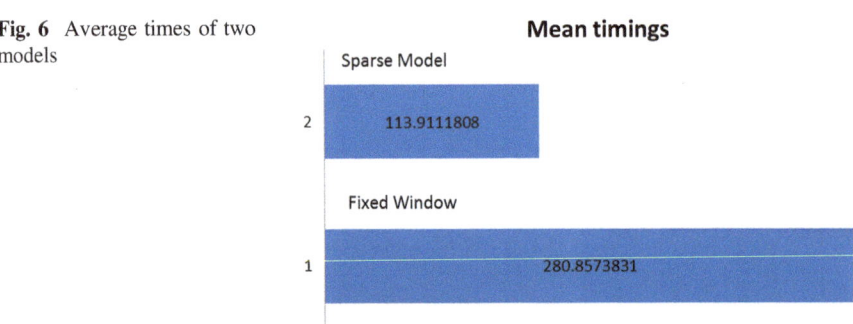

region. For example, assume that the observed minimum and maximum values of
x and y coordinates of segmented portion in sequence are 10, 20, 140 and 220.
Then, the fixed window is from 10×20 to 140×220. The time taken by the 100
mammograms for the construction of GLAM is reported in Table 1. After experimentation, in average, sparse model completes the construction of GLAM in
113.911 s whereas time taken by the choice of fixed window for ROI is 280.86 s. In
specific, the average time taken by sparse-modeled ROI for the selected images in
Fig. 1 is 10.5, 434, 072 and 975 s whereas the existing rectangular-shaped ROI is
328.2470765 s. Figure 5 represents the timings of the two models for 100 mammograms, and Fig. 6 depicts the mean time taken by the two models for 100
mammograms. Results have shown that existing fixed window model takes more
than double the time than proposed sparse model.

4.1 Limitation

This method suffers from the disadvantage of taking more memory whenever tissue region is comparatively high. Though consuming high memory is depicted as the limitation in reality, it is not so as on the present day computing systems memory space is not at all is a constraint.

5 Conclusion

Existing methods based on regular-shaped ROI works well when the tissue portion occupies 80–90 % of the selected rectangular area of ROI. The above works are silent about their performance whenever the tissue area is comparatively low in selected ROI. The procedure proposed in this paper performs extraordinary well in drastically reducing time complexity whenever the mass is of arbitrary shaped and small in size. Results are shown that existing fixed window model takes more than double the time than ours.

Studying the performance of sparse-modeled ROI in various other gray level matrices and on other applications like face detection is our future endeavor.

Acknowledgments This work is funded by Department of Science and Technology, New Delhi, India.

References

1. Winsberg F, Elkin M, Macy J, Bordaz V, Weymouth W (1967) Detection of radiographic abnormalities in mammograms by means of optical scanning and computer analysis. Radiology 89:211–215
2. Bozek J, Mustra M, Delac K, Grgic M (2009) A survey of image processing algorithms in digital mammography. In: Grgic et al (eds) Recent advances in multimedia signal processing and communications, SCI 231, pp 631–657
3. Pisano ED, Cole EB, Hemminger BM, Yaffe MJ, Aylward SR, Maidment ADA, Eugene Johnston R, Williams MB, Niklason LT, Conant EF, Fajardo LL, Kopans DB, Brown ME, Pizer SM (2000) Image processing algorithms for digital mammography: a pictorial essay. J Radiogr 20(5):400–420
4. Mohanty AK, Senapati MR, Lenka SK (2013) A novel image mining technique for classification of mammograms using hybrid feature selection. Neural Comput Appl 22:1151–1161
5. Haralick RM, Shanmugam K, Dinstein I (1973) Textural features for image classification. IEEE Trans Syst Man Cybern 3(6):610–621
6. Wiesmuller S, Chandy DA (2010) Content based mammogram retrieval using Gray Level Aura Matrix. Int J Comput Commun Inf Syst (IJCCIS) 2(1):217–223
7. Qin X, Yang Y-H (2004) Similarity measure and learning with gray level aura matrices (GLAM) for texture image retrieval. In: Proceedings of the 2004 IEEE computer society conference on computer vision and pattern recognition (CVPR'04), vol 4. IEEE

8. Chang HD, Shi XJ, Min R, Hu LM, Cai XP, Du HN (2006) Approaches for automated detection and classification of masses in mammograms. Pattern Recogn 39:646–668
9. Ke L, Mu N, Kang Y (2010) Mass computer-aided diagnosis method in mammogram based on texture features. In: Biomedical engineering and informatics (BMEI), 3rd international conference. IEEE explore, pp 146–149
10. Jalja K, Bhagvati C, Deekshatulu BL, Pujari AK (2005) Texture element feature characterizations for CBIR. In: Proceedings of geoscience and remote sensing symposium (IGARSS '05), vol 2
11. Choraś RS (2008) Feature extraction for classification and retrieval mammogram in databases. Int J Med Eng Inf 1(1):50–61
12. Khuzi AM, Besar R, Wan Zaki WMD (2008) Texture features selection for masses detection in digital mammogram. In: 4th Kuala Lumpur international conference on biomedical engineering, IFMBE proceedings, vol 21. part 3, part 8, pp 629–632
13. Haliche Zohra, Hammouche Kamal (2011) The gray level aura matrices for textured image segmentation. Analog Integr Circ Sig Process 69:29–38
14. Chandy DA, Johnson JS, Selvan SE (2014) Texture feature extraction using gray level statistical matrix for content-based mammogram retrieval. Multimed Tools Appl 72 (2):2011–2024
15. Mohanty AK, Senapati MR, Beberta S, Lenka SK (2013) Texture-based features for classification of mammograms using decision tree. Neural Comput Appl 23:1011–1017
16. Mohanty AK, Senapati MR, Lenka SR (2013) An improved data mining technique for classification and detection of breast cancer from mammograms. Neural Comput Appl 22 (1):303–310
17. Hussain M (2014) False-positive reduction in mammography using multiscale spatial Weber law descriptor and support vector machines. Neural Comput Appl 25(1):83–93
18. Mohanty AK, Senapati MR, Beberta S, Lenka SK (2013) Mass classification method in mammograms using correlated association rule mining. Neural Comput Appl 23:273–281

A Survey on Identification of Protein Complexes in Protein–protein Interaction Data: Methods and Evaluation

Praveen Tumuluru, Bhramaramba Ravi and Sujatha Ch

Abstract Since identification of protein complexes from protein–protein interaction (PPI) networks plays an important role in the computational biology, in this paper, we discuss different types of protein complex identification algorithms such as Markov Clustering algorithm, ClusterBFS, Connected Affinity Clique Extension, PE-weighted Clustering algorithm, Detection of Protein Complex Core and Attachment Algorithm and Dynamic Protein Complex Algorithm. Thereafter, we focus on computational analysis of protein complexes through various measures and various protein interaction databases, with which we can detect protein complexes effectively and efficiently.

Keywords Protein–protein interaction network · Protein complex · Computational biology

1 Introduction

A protein complex is a group of associated polypeptide chains linked by non-covalent protein–protein interactions (PPIs). PPI networks play an important role in the biological processes including cell cycle control, discrimination, protein folding, signal transduction, transcription, and translation [1].

P. Tumuluru (✉)
Department of Computer Science and Engineering, GITAM University,
Visakhapatnam, India
e-mail: praveenluru@gmail.com

B. Ravi
Department of Information Technology, GITAM University, Visakhapatnam, India
e-mail: bhramarambaravi@gmail.com

S. Ch
Department of Computer Science and Engineering, Acharya Nagarjuna University,
Guntur, India
e-mail: suzychopra@gmail.com

© The Author(s) 2015
N.B. Muppalaneni and V.K. Gunjan (eds.), *Computational Intelligence Techniques for Comparative Genomics*, Forensic and Medical Bioinformatics,
DOI 10.1007/978-981-287-338-5_5

Detecting the protein complexes from the available PPI data will help to deeply understand biological activity mechanism and the architecture of protein interaction network (PIN). In recent years, a huge number of computational approaches based on graph clustering have been applied to PPI networks for protein complexes identification. These graph clustering algorithms mainly depend on the association of topological analysis of PPI networks to classify protein complexes [2–5].

Computational approaches can be applied to identify protein complex information by searching closely connected regions in a PPI network [3], which is like a graphical map of an entire organism's interactome. This is assembled from existing PPI knowledge by considering unit of proteins as nodes and the subsistence of physical interactions between a pair of proteins as connections. The titanic quantity of genes and proteins that participate in biological networks that inflict the need for determination of protein complexes within the network, while these complexes will be the first step in decoding the composite genetic or cellular interactions of the overall network. Many algorithms for detecting protein complexes from PPI network have been developed, and these algorithms adopt different strategies to detect protein complexes and thus obtain different results.

2 Review of Algorithms

In this section, we first pioneer some basic terminologies for graphs and then review the different algorithms for protein complex detection.

2.1 Terminology

Given a PPI network $G = (V, E)$ with a set of nodes $V = \{v_1, v_2, \ldots, v_n\}$ and a set of edges $E = \{e_1, e_2, \ldots, e_m\}$, where the nodes represent proteins and edges represent pairwise interactions. A walk is a sequence of vertices where edges exist between two adjacent vertices. The neighborhood of a vertex v in a graph G is the induced subgraph of G consisting of all vertices adjacent to v and all edges connecting two such vertices. A k-core is a subgraph in which all the vertices have degrees no less than k and the order of a k-core is k, if it is not a $(k + 1)$-core. Given two graphs $A = (V_A, E_A)$ and $B = (V_B, E_B)$, their neighborhood affinity NA(A, B) is defined as to measure the similarity between them.

$$\mathrm{NA}(A, B) = \frac{|V_A \cap V_B|^2}{|V_A| \times |V_B|} \tag{1}$$

For a given PPI network $G = (V, E)$, the degree of a vertex $v \in V$ is the number of v's neighbors in G, represented as deg(v). The average degree of graph G is defined as the average of deg(u) for all $u \in V$, represented as Avdeg(G). The density of a graph G denoted as den(G),

$$Avdeg(G) = \sum_{u \in V} \frac{\deg(u)}{|V|} \tag{2}$$

$$den(G) = \frac{2 \times |E|}{|V| \times (|V| - 1)} \tag{3}$$

2.2 Algorithms

2.2.1 Markov Clustering Algorithm (MCL)

The MCL algorithm simulates random walks within a graph by alternation of two operators called expansion and inflation. Expansion coincides with taking the power of a stochastic matrix using the normal matrix product (i.e., matrix squaring).

Inflation corresponds with taking the Hadamard power of a matrix, followed by a scaling step, such that the resulting matrix is stochastic again, i.e., the matrix elements (on each column) correspond to probability values [6].

Expansion and inflation represent different tidal forces which are alternated until an equilibrium state is reached. An equilibrium state takes the form of a so-called doubly idempotent matrix, i.e., a matrix that does not change with further expansion of inflation steps. At each step, the algorithm finds the minimal cut which allows separating the graph component into two clusters by minimizing the number of inter-cluster edges. The process is repeated until a stop condition is reached.

2.2.2 Cluster Breadth-first Search

Given a weighted network, the objective of this algorithm is to output a set of disjoint dense subgraphs. Model the network as an undirected graph $G = (V, E)$ with a confidence score $0 < W_{u,v} \leq 1$, for every edge $(u, v) \in E$. For any pair of vertices, u and v without an edge between them, we set $W_{u,v} = 0$. For each set of vertices $S \subset E$, define its weighted density as the sum of the weights of the edges between them divided by the total number of possible edges. In other words, the density of a set is measure of how close the induced subgraph is to clique, varies from 0 to 1.

$$D_\omega(S) = \frac{\sum_{(u,v) \in S} \omega_{u,v}}{|S| * (|S| - 1)/2} \tag{4}$$

ClusterBFS assembles one cluster at a time, and every cluster is expanded from a unique seed protein. The unclustered node is added if it has the highest edge weight, and the density of the cluster remains higher than a user-defined threshold T_s; otherwise, the cluster is output. Consequently, the ClusterBFS has two parameters: T_s, the weighted density threshold and R.

In the PPI network, each vertex is represented as seed. After achieving the seed vertex, we use the breadth-first search technique to produce each cluster in terms of the weighted density. For a cluster, at each step, we have a current vertex set C, which primarily contains one seed as protein v. Then, from all unclustered vertices, we search for the vertex u with maximum value of the edge weight that is adjacent to the v in BFS [7].

If the weighted density of the cluster is smaller than a threshold, we stop intensifying this cluster and output it. If not, we put vertex u into C and update the density value. If the density value is smaller than our density threshold T_s, we do not include u in the cluster and output C. We repeat this procedure until all vertices in the graph are clustered.

Since all vertices in the graph have been selected as seeds, the clusters produced have large overlaps, which will result in high redundancy. Hence, a redundancy-filtering procedure is designed to process candidate clusters and finally generate protein complexes by eliminating such kind of redundancy.

2.2.3 Connected Affinity Clique Extension (CACE)

In this algorithm, the value of the protein connected affinity which is inferred from protein complexes is interpreted as the reliability and possibility of interaction.

The algorithm works by first detecting a set of seeds in the graph G, then expanding these seeds into modules based on affinity clique extension. Firstly, the input as PIN and PCT information to identified the protein complexes modules [8].

Step: (1) The process starts to utilize the available data for building the Weighted PIN (protein interaction dataset). Calculate the connected affinity coefficient value for each complex from protein complex dataset represented as CAC_{ij} and also get the value of interaction for each pair of proteins from PPI network represented as PB_{ij} [9].

Based on the above value which affects the protein relations and then calculates the weight to represent the likelihood interaction between any two proteins:

$$WP_iP_j = \alpha CAC_{ij} + (1 - \alpha)\, PB_{ij},$$

where α is a parameter.

Now, we can construct a Weighted PIN using the above values.

Step: (2) Now, we choose the seeds from the weighted PIN by finding all of the maximal cliques with size greater than CZ in the network [10].

Step: (3) Now, mining the complexes from the connected affinity clique extension (CACE) which is the key point to determine the quality of community.

From the number of cliques or seeds of the communities are obtained and to get the final community the seed has to expand until all the satisfactory nodes are included. For the final community judgment set a threshold parameter T_d, and utilizing it to decide which node is to be added into it [11–14].

Finally, the algorithm shows that the CACE can detect the functional modules much more effectively and accurately when compared with other algorithms.

2.2.4 PE-weighted Clustering Coefficient (PE-WCC)

In this algorithm, it first assesses the reliability of the protein interaction data using the PE measure and then predicts the protein complexes based on the concept of weighted clustering coefficient. It detects the more matched complexes with higher quality scores.

Step: (1) For assessing the reliability of the protein interaction data, we introduce the value called as PE measure where it reduces the noise level associated with the PPI networks

$$(p_k)_{ij} = 1 - \prod_{v_l} (1 - (p_{k-1})_{il} \cdot (p_{k-1})_{jl}) \tag{5}$$

where the product by all v_l: $(v_l, v_i) \in E$, $(v_l, v_i) \in E$ then for each protein in the PPI network, we calculate the average PE measures.

$$(W_{\text{avg}})_i = \frac{\sum_{v_l} p_{il}}{N_i} \tag{6}$$

where v_l: $(v_l, v_i) \in E$, N_i is the number of the neighbors of v_i and $i = 1, \ldots, N$. If the PE measure p_{il} is less than the average $(w_{\text{avg}})_i$, then edge between proteins i and l is considered unreliable and therefore, it should be removed from the network.

Step: (2) For detecting protein complex using weighted clustering coefficient for each protein v_i in the PPI network, we first create the neighborhood graph, calculate the weighted clustering coefficient, and then calculate the degree of each node in the neighborhood graph; the "degree" of a node being the number of its neighbors. The weighted clustering coefficient c_i in this case is calculated according to the following formula:

$$c_i = \frac{2 \cdot N_{3\text{cliques}}}{N_i^2 \cdot (N_i - 1)} \tag{7}$$

where N_3 cliques are the number of 3 cliques in the neighborhood graph. Once the degree is calculated, we sort the sequence of proteins in the neighborhood graph accordingly from minimum to maximum [15]. The protein v_j with the lowest degree and its corresponding interactions are removed from the neighborhood graph and c_i is recalculated. This process stops when the neighborhood graph contains only 3 proteins and the sequence of proteins with the highest c_i is returned as a valid core protein complex.

Assessing the quality of predicted complexes to evaluate the accuracy of the proposed method, we used the Jaccard index which defined as follows:

$$\text{MathScore}(K, R) = \frac{|V_K \cap V_R|}{|V_K \cup V_R|} \tag{8}$$

where K is a cluster and R is a reference complex. V_K and V_R are the set of proteins in K and R, respectively. The complex K is defined to match the complex R if match score $(K, R) \geq \alpha$.

2.2.5 Detection of Protein Complex Core and Attachment Algorithm

The basic idea behind this method resides of two main steps. In the first step, it extracts the complex core based on the weighted *PPI* network from the view of edge. In the second step, it identifies the attachments for each core to form the protein complex.

The protein complex core is defined as follows:

1. Each complex has a set of core proteins.
2. The core has relatively more interactions; the functional similarity between core proteins should be high [16].

Detecting the Protein Complex Core

The *PPI* interaction data are an undirected weighted graph $G = (V, E, W)$ where V represents the proteins and E represents the interactions between proteins and W is a weight to each edge with co-expressed correlation between proteins. The sum of weight is defined as,

$$|W| = \sum_i^{|V|} \sum_j^{|V|} A_{ij} \tag{9}$$

The average weight of graph G and the weighted density of the core are calculated with Eqs. (2) and (3). The neighborhood affinity for the two graphs and to measure the similarity between the graphs, it filters the redundant core and determines whether the predicted protein complex matches the real protein complex by using the Eq. (1).

Next, the method considers the pairwise proteins with high affinity score as core proteins and forms protein complex core from edge. Firstly, the preliminary cores are detected from the neighborhood graph of the pairwise proteins in the PPI network. If the neighborhood graph satisfies the requirement, it is directly regarded as preliminary core, otherwise, will be dealt with until it meets the requirements. Then, due to different pairwise proteins may produce same or similar preliminary cores, which will lead to high redundancy. In order to eliminate the redundant cores, we use redundancy filtering procedure to process preliminary cores and finally get the protein complex cores.

Detecting the Attachment

It identifies the attachment proteins for each core to form protein complex. In this method, it identifies attachment proteins by edge [17]. The criteria for selecting attachment protein not only apply the properties of network topology but also consider the internal affinity between core and attachment.

Generally, the attachment proteins densely interact with core proteins than other attachment proteins. The affinity score between core and attachment is high [16]. It may participate in multiple complexes unlike the unique core. To identify such proteins, a function is needed to measure the closeness between attachment and core.

$$\text{closeness}(v, \text{CG}) = \frac{|N_v \cap V_{\text{CG}}|}{|V_{\text{CG}}|} \tag{10}$$

Using this method, those proteins closely associated with complex core can be selected as attachment proteins. However, there exists many attachment proteins, they only interact with a few core proteins, but the gene co-expressed consistency and the function similarity is high. The average weight between attachment proteins and core proteins is defined as,

$$\text{AW}(v, \text{CG}) = \frac{|W_C|}{|N_V \cap V_{\text{CG}}|} \tag{11}$$

Finally, in this method, it identifies the attachment proteins for each core to form protein complex.

2.2.6 Dynamic Protein Complex Algorithm

The algorithm DPC operates in three stages:

1. detecting protein complex cores,
2. generating potential dynamic complexes, and
3. filtering false-positive complexes.

Detecting Protein Complex Cores

Detecting possible protein complex cores is divided into two main steps: searching always active proteins and forming possible protein complex cores. Forming possible protein complex cores is designed to group always active proteins into many connected subgraphs according to the topological and dynamic features of protein complex cores. The core candidate proteins are sorted by the numbers of their own always active neighbors in descending order in G [18]. The sorted always active proteins will be stored into a queue Q. The first vertex in the queue Q is picked and used as a seed to grow a new possible protein complex core. Once a protein complex core is completed, all vertices in it will be tagged with "1" and cannot be extended into any other protein complex cores [19].

A final protein complex core is generated by removing vertices recursively from the preliminary core according to the edge clustering coefficient until its density is larger than or equal to a given threshold T_d. For an edge$(u, v) \in E$, its edge clustering coefficient [16] $\mathrm{ECC}(v, u)$ is defined as the number of triangles to which (u, v) belongs and is divided by the number of triangles that might potentially include (u, v), as

$$\mathrm{ECC}(u, v) = \frac{Z_{(u,v)}}{\min\{\deg(u) - 1, \ \deg(v) - 1\}} \quad (12)$$

For a preliminary core C with $\mathrm{den}(C) < T_d$, the edge clustering coefficient ECC (u, v) of each edge connecting the seed vertex v and a rest vertex u is calculated. Then, the vertex with minimum ECC will be removed from the preliminary core C [20]. Suppose, if the preliminary core C with $\mathrm{den}(C) \leq T_d$ is outputted as a potential protein complex core.

Generating Potential Dynamic Protein Complexes

After finding the potential protein complex cores, the important point is how to find the attachments for each core to form protein complexes. For this, we first find a best protein complex core for each potential attachment in the set. We say that there is a connection between a protein complex core and an attachment if there is at least an edge connecting the attachment and one protein in the protein complex core. For each attachment in the set, we select a best one with the highest value of "closeness" as its potential protein complex core. The "closeness" of an attachment V to a core C_i is

$$CL(V, C_i) = \sum_{u \in C_l} ECC(v, u) \tag{13}$$

If the "closeness" of an attachment to all the possible protein complex cores is equal to 0, this attachment will be ignored. For each time progress in the molecular cycle, every protein will be judged whether it is active or not. If a protein is active in a certain time course, it will be considered as an attachment and added to its corresponding protein complex core at this time course. Finally, the protein complex cores and their attachments with the expressed time course will be outputted.

Filtering False-positive Complexes

In the last stage, it will reexamine the possible dynamic protein complexes to filter some false positives. According to the formation and function of a protein complex, it should be active in two or more continual time courses. It uses some of the rules to filter false-positive complexes.

1. A protein complex should include at least two proteins.
2. The attachment proteins should be active in the same time course or in different but adjacent time courses.
3. If the attachments of a possible protein complex do not satisfy the second rule and the protein complex core involves at least two proteins, the core will be kept as a final protein complex.

Finally, one can identify the protein complex with these stages, namely detecting protein complex cores, generating potential dynamic complexes, and filtering false-positive complexes.

3 Computational Analysis of Protein Complexes

To evaluate the clustering techniques against explicit datasets that previously contain information about recorded protein complexes [21]. For comprehensive evaluation, we employed several valuation measures with "co-annotation, co-localization, precision, recall, F-measure, sensitivity, positive predictive value, and separation values."

3.1 Co-annotation and Co-localization Within Predicted Complexes

Since protein complexes are created to achieve a specific cellular function, proteins within the same complex tend to allocate common functions and be co-localized. Generally, higher co-annotation and co-localization scores [22] show that proteins within the same protein complexes tend to share higher functional similarity, and hence, they can be used to evaluate the overall quality of predicted protein complexes. The matching measures should be auxiliary with scores that evaluate the biological consequence of predicted complexes supported on the co-localization and co-annotation of the necessary proteins instead of relying on predefined standard protein complexes. The "biological process" and "cellular component" categorization are used to compute the co-annotation score and co-localization score correspondingly.

3.2 Precision, Recall, and F-measure

The neighborhood affinity score between a predicted complex p and a real complex b in the benchmark, $NA(p, b)$, is used to determine whether they match with each other. If $NA(p, b) \geq \omega$, they are considered to be matching. We assume that P and B are the set of complexes predicted by a computational method and real ones in the benchmark, respectively. N_{cp} is the number of predicted complexes which match at least a real complex and N_{cb} is the number of real complexes that match at least a predicted one [21]. Precision is a measure of exactness and recall is a measure of completeness and is defined as,

$$
\begin{aligned}
N_{cp} &= |\{p|p \in P, \exists b \in B, NA(p,b) \geq \omega\}|, \\
N_{cb} &= |\{b|b \in B, \exists p \in P, NA(p,b) \geq \omega\}|,
\end{aligned}
\tag{14}
$$

$$
\text{Precision} = \frac{N_{cp}}{|P|} \text{ and Recall} = \frac{N_{cb}}{|B|}
\tag{15}
$$

F-score—measures test accuracy, weighted average of precision and recall (in [0, 1]), and F-measure, as the harmonic mean of precision and recall—can be used to evaluate the overall performance of the different techniques,

$$
F-\text{measure} = 2 \cdot \frac{\text{precision} \cdot \text{recall}}{\text{precision} + \text{recall}}
\tag{16}
$$

3.3 Sensitivity, Positive Predictive Value and Separation Values

Sensitivity (S_n) and positive predictive value (PPV) are also used to evaluate the accuracy of the predictions [3, 23]. Given n benchmark complexes and m predicted complexes, i-th is the number of proteins in common between its benchmark complex and j-th predicted complex. S_n and PPV are defined as:

$$S_n = \frac{\sum_{i=1}^{n} \max_{j}\{T_{ij}\}}{\sum_{i=1}^{n} N_i} \text{ and } PPV = \frac{\sum_{j=1}^{m} \max_{i}\{T_{ij}\}}{\sum_{j=1}^{m} T_j} \qquad (17)$$

Here, N_i is the number of proteins in the i-th benchmark complex, and $T_j = \sum T_{ij}$.

Generally, high S_n values show that the prediction has a good coverage of the proteins in the real complexes and high PPV values specify predicted complexes are probably to be true positive. The accuracy of a prediction, Acc, is finally defined as the geometric average of sensitivity and positive predictive value,

$$Acc = \sqrt{S_n \times PPV}. \qquad (18)$$

A new measure called separation [3] is proposed to emphasize the one-to-one association between a predicted complex and a real complex. The separation value for the i-th benchmark complex and j-th predicted complex, sep_{ij}, is defined as:

$$sep_{ij} = \frac{T_{ij}}{\sum_{i=1}^{n} T_{ij}} \times \frac{T_{ij}}{\sum_{j=1}^{m} T_{ij}}. \qquad (19)$$

The complex-wise separation sep_{ij} and the cluster-wise separation sep_p are defined in Eq. (8). The final geometrical separation value (sep) is defined as the geometrical mean of sep_{ij} and sep_p

$$sep_{ij} = \frac{\sum_{i=1}^{n} \sum_{j=1}^{m} sep_{ij}}{n} \text{ and } sep_p = \frac{\sum_{j=1}^{n} \sum_{i=1}^{m} sep_{ij}}{m} \qquad (20)$$

4 Protein Interaction Databases

The massive quantity of experimental PPI data being produced on steady basis has lead to the construction of computer readable biological databases in order to organize and to process this data. For example, the biomolecular interaction network database (BIND) is created on an extensible specification system that permits

Table 1 Protein interaction databases

S. No	Database name	Total no. of interactions	References	Source link	Number of species/ organisms
1	BioGrid	7,17,604	Stark et al. [36]	http://thebiogrid.org/	60
2	DIP	76,570	Xenarios et al. [37]	http://dip.doe-mbi.ucla.edu/ dip/Main.cgi/	637
3	Hit predict	2,39,584	Patil et al. [38]	http://hintdb.hgc.jp/htp/	9
4	MINT	2,41,458	Chatr-aryamontri et al. [39]	http://mint.bio.uniroma2.it/ mint/	30
5	IntAct	4,33,135	Hermjakob et al. [40]	http://www.ebi.ac.uk/intact/	8
6	APID	3,22,579	Prirto and de Las Rivas [41]	http://bioinfow.dep.usal.es/ apid/index.html/	15
7	BIND	3,00,000+	Bader et al. [42]	http://bind.ca/	–
8	PINA 2.0	3,00,155	Cowley et al. [43]	http://cbg.garvan.unsw.edu. au/pina/	7

an elaborate description of the manner in which the PPI data were derived experimentally, often including links directly to the concluding evidence from the literature [24]. The database of interacting proteins (DIP) is another database of experimentally determined protein–protein binary interactions [25]. The biological general repository for interaction datasets (BioGRID) is a database that contains protein and genetic interactions among thirteen different species [26]. Interactions are regularly added through exhaustive curation of the primary literature to the databases. Interaction data are extracted from other databases including BIND and Munich Information Center for protein sequences (MIPS) [27], as well as directly from large-scale experiments [28]. Hit predict is a resource of high confidence PPIs from which we can get the total number of interactions in a species for a protein and can view all the interactions with confidence scores [29].

The Molecular INTeraction (MINT) database is another database of experimentally derived PPI data extracted from the literature, with the added element of providing the weight of evidence for each interaction [25]. The Human Protein Interaction Database (HPID) was designed to provide human protein interaction information precomputed from existing structural and experimental data [30]. The Information Hyperlinked over Proteins (iHOP) database can be searched to identify previously reported interactions in PubMed for a protein of interest [31]. IntAct [32] provides an open source database and toolkit for the storage, presentation, and analysis of protein interactions [33].

The web interface provides both textual and graphical representations of protein interactions and allows exploring interaction networks in the context of the GO annotations of the interacting proteins [24]. However, we have observed that the intersection and overlap between these source PPI databases are relatively small. Recently, the integration has been done and can be explored in the web server called Agile Protein Interaction Data Analyzer (APID) which is an interactive bioinformatics' web tool developed to allow exploration and analysis of currently known information about PPIs integrated and unified in a common and comparative platform [34].

The Protein Interaction Network Analysis (PINA2.0) platform is a comprehensive web resource, which includes a database of unified PPI data integrated from six manually curated public databases and a set of built-in tools for network construction, filtering, analysis, and visualization [35, 44]. The databases and number of interactions were tabled in Table 1.

5 Conclusion

In this paper, we have discussed different algorithms to detect protein complexes in large interaction networks. Protein complexes are significant for recognizing the principles of cellular organization and function. The motivation behind these algorithms is to benchmark the clustering techniques and measure their prediction accuracy to identify the protein complexes. The clustering result of each computational method can be regarded as a feature that describes the PPI network. However, most of the approaches rely on the hypothesis that proteins within the same complex would have relatively more interactions. Protein complexes computation can be measured efficiently with the help of precision, recall, F-measure, co-localization and co-annotation, sensitivity, positive predictive value, and separation values, etc. With the help of this protein complex identification, one can detect the behavior and closeness of proteins and which will helpful in drug detection. The authors declare that there is no conflict of interests regarding the publication of this paper.

Acknowledgement The authors published this paper under their Ph.D. work. The authors wish to thank the University Grants Commission (UGC) for extending financial support for this study, under the project "Development of a Software Tool to Identify Lung-Cancer Related Genes using Protein-Protein Interaction Network" with sanction F.NO:4-4/2014-15[MRP-SEM/UGC-SERO].

References

1. Tu S, Xu L (2010) A binary matrix factorization algorithm for protein complex prediction. In: IEEE international conference on bioinformatics and biomedicine workshops (BIBMW), ISBN: 978-1-4244-8304-4
2. Ou-Yang L, Dai DQ, Zhang XF (2013) Protein complex detection via weighted ensemble clustering based on Bayesian nonnegative matrix factorization. PLoS One. doi:10.1371/journal.pone.0062158

3. Li X, Wu M, Kwoh C, Ng S (2010) Computational approaches for detecting protein complexes from protein interaction networks. BMC Genom 11:S3. doi:10.1186/1471-2164-11-s1-s3

4. Ji J, Zhang A, Liu C, Quan X, Liu Z (2012) Functional module detection from protein-protein interaction networks. IEEE Trans Knowl Data Eng 1

5. Brohee S, Van Helden J (2006) Evaluation of clustering algorithms for protein-protein interaction networks. BMC Bioinformatics 7:488. doi:10.1186/1471-2105-7-488

6. van Dongen S (2000)Graph clustering by flow simulation. Ph. D. thesis. University of Utrecht, The Netherlands, May 2000

7. Tang X, Wang J, Li M, He Y, Pan Y (2014) A novel algorithm for detecting protein complexes with the breadth first search. BioMed Res Int 2014:354539

8. Adamcsek B, Palla G, Farkas IJ et al (2006) CFinder: locating cliques and overlapping modules in biological networks. Bioinformatics 22:1021–1023

9. Baumes MGJ, Magdon-Ismail M (2004) Efficient identification of overlapping communities. Intelligence and security informatics, vol 3495. Springer, Berlin, pp 27–36

10. Bron C, Kerbosch J (1973) Algorithm 457: finding all cliques of an undirected graph. Commun ACM 16(9):575–577

11. Lee C, Reid F, McDaid A et al. (2010) Detecting highly overlapping community structure by greedy clique expansion. Physics, pp 1002–1827

12. Zhang S, Ning X, Zhang XS (2006) Identification of functional modules in a PPI network by clique percolation clustering. Comput Biol Chem 30(6):445–451

13. Wang J, Liu B, Li M et al (2010) Identifying protein complexes from interaction networks based on clique percolation and distance restriction. BMC Genomics 11(Suppl 2):S10

14. Qi Y, Balem F, Faloutsos C et al (2008) Protein complex identification by supervised graph local clustering. Bioinformatics 24(13):i250–i268

15. Zaki N, Efimov D, Berengueres J (2013) Protein complex detection using interaction reliability assessment and weighted clustering coefficient. BMC Bioinformatics. doi:10.1186/1471-2105-14-163

16. Gavin AC, Superti-Furga G (2003) Protein complexes and proteome organization from yeast to man. Curr Opin Chem Biol 7(1):21–27

17. Zhao J, Hu X, He T, Li P, Zhang M, Shen X (2014) An edge-based protein complex identification algorithm with gene co-expression data (PCIA-GeCo). IEEE Trans Nanobiosci 13(2). ISSN: 1536-1241

18. Li M, Chen W, Wang J, Wu F-X, Pan Y (2004) Identifying dynamic protein complexes based on gene expression profiles and PPI networks. BioMed Res Int (Article ID 375262)

19. Wang J, Li M, Deng Y, Pan Y (2010) Recent advances in clustering methods for protein interaction networks. BMC Genomics 11(suppl 3):S10

20. Wang J, Peng X, Peng W, Wu F (2014) Dynamic protein interaction network construction and applications. Proteomics 14(4–5):338–352

21. Wu M, Li X, Kwoh CK, Ng SK (2009) A core-attachment based method to detect protein complexes in PPI networks. BMC Bioinformatics 10:169. doi:10.1186/1471-2105-10-169

22. Friedel CC, Krumsiek J, Zimmer R (2008) Bootstrapping the interactome: unsupervised identification of protein complexes in yeast. In: Vingron M, Wong L (eds) 12th Annual international conference on research in computational molecular biology (RECOMB), of LNCS, vol 4955. Springer, Berlin, pp 3–16

23. Li P, He X, Hu X, Zhao J, Shen X, Zhang M, Wang Y (2014) A novel protein complex identification algorithm based on connected affinity clique extension (CACE). IEEE Trans nanobiosci 13:89. ISSN: 1536-1241

24. Srinivasa Rao V, Srinivas K, Sujini GN, Sunand Kumar GN (2014) Review article protein-protein interaction detection: methods and analysis. Int J Proteomics, vol 2014, Article ID 147648

25. Xenarios I, Salwínski Ł, Duan XJ, Higney P, Kim S-M, Eisenberg D (2002) DIP, the database of interacting proteins: a research tool for studying cellular networks of protein interactions. Nucleic Acids Res 30(1):303–305

26. Stark C, Breitkreutz B-J, Reguly T, Boucher L, Breitkreutz A, Tyers M (2006) BioGRID: a general repository for interaction datasets. Nucleic Acids Res 34:D535–539
27. Mewes HW, Ruepp A, Theis F et al (2011) MIPS: curated databases and comprehensive secondary data resources in 2010. Nucleic Acids Res 39(1):D220–D224
28. Chatr-aryamontri A, Ceol A, Palazzi LM et al (2007) MINT: the molecular interaction database. Nucleic Acids Res 35(1):D572–D574
29. Patil A, Nakai K, Nakamura H (2011) HitPredict: a database of quality assessed protein-protein interactions in nine species. Nucleic Acids Res 39(1):D744–D749
30. Han K, Park B, Kim H, Hong J, Park J (2004) HPID: the human protein interaction. Bioinformatics 20(15):2466–2470
31. Fernández JM, Hoffmann R, Valencia A (2007) iHOP web services. Nucleic Acids Res 35: W21–W26
32. Hermjakob H, Montecchi-Palazzi L, Lewington C et al (2004) IntAct: an open source molecular interaction database. Nucleic Acids Res 32:D452–D455
33. Hermjakob H, Montecchi-Palazzi L, Lewington C, Mudali S, Kerrien S, Orchard S, Vingron M, Roechert B, Roepstorff P, Valencia A, Margalit H, Armstrong J, Bairoch A, Cesareni G, Sherman D, Apweiler R (2004) IntAct: an open source molecular interaction database. Nucleic Acids Res 32(Database issue):D452–D455
34. Prieto C, de Las Rivas J (2006) APID: agile protein interaction dataanalyzer. Nucleic Acids Res 34:W298–W302
35. Cowley MJ, Pinese M, Kassahn KS et al (2012) PINA v2. 0: mining interactome modules. Nucleic Acids Res 40:D862–D865
36. Stark C, Breitkreutz B-J, Reguly T, Boucher L, Breitkreutz A, Tyers M (2006) BioGRID: a general repository for interaction datasets. Nucleic Acids Res 34:D535–539
37. Xenarios I, Salwinski L, Duan XJ, Higney P, Kim S-M, Eisenberg D (2002) DIP, the database of interacting proteins: a research tool for studying cellular networks of protein interactions. Nucleic Acids Res 30(1):303–305
38. Patil A, Nakai K, Nakamura H (2011) HitPredict: a database of quality assessed protein-protein interactions in nine species. Nucleic Acids Res 39(1):D744–D749
39. Chatr-aryamontri A, Ceol A, Palazzi LM et al (2007) MINT: the Molecular interaction database. Nucleic Acids Res 35(1):D572–D574
40. Hermjakob H, Montecchi-Palazzi L, Lewington C et al (2004) IntAct: an open source molecular interaction database. Nucleic Acids Res 32:D452–D455
41. Prieto C, de Las Rivas J (2006) APID: agile protein interaction dataanalyzer. Nucleic Acids Res 34:W298–W302
42. Bader GD, Donaldson I, Wolting C, Ouellette BFF, Pawson T, Hogue CWV (2001) BIND: the bimolecular interaction network database. Nucleic Acids Res 29(1):242–245
43. Cowley MJ, Pinese M, Kassahn KS et al (2012) PINA v2. 0: mining interactome modules. Nucleic Acids Res 40:D862–D865
44. Dezso Z, Oltvai ZN, Barabasi A-L (2003) Bioinformatics analysis of experimentally determined protein complexes in the yeast Saccharomyces cerevisiae. Genome Res 13 (11):2450–2454

Author Biographies

Praveen Tumuluru received the M. Tech degree in Computer Science and Engineering from Koneru Lakshmaiah College of Engineering, Acharya Nagarjuna University, in 2008. He is a research Scholar in GITAM University working in Data Mining Techniques for Bioinformatics Protein–protein Interaction. He has been working as Assistant Professor, in the Department of Electronics & Computer Engineering, Prasad V. Potluri Siddhartha Institute of Technology, Vijayawada, since 2008.

Bhramaramba Ravi presently working as Associate Professor in GITAM University. She has a total of 9 years of research experience and 14 years of teaching. She received her Ph.D. from Jawaharlal Nehru Technological University, in 2011 and MS degree in Software Systems from BITS, in 1999. She has published 13 papers in various National and International Journals/Conferences. Her current research interests are in the areas of Data Mining Techniques for Bioinformatics.

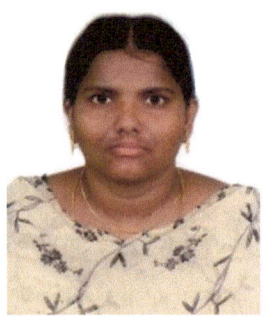

Sujatha Ch received the M.Tech degree in Computer Science and Engineering from NEC College of Engineering, Jawaharlal Nehru Technological University, in 2009. She is a research Scholar in Acharya Nagarjuna University working in Data Mining Techniques for Bioinformatics Protein–protein Interaction. She has been working as Assistant Professor, in the Department of Electronics & Computer Engineering, Prasad V. Potluri Siddhartha Institute of Technology, Vijayawada, since 2011.

Modeling Artificial Life: A Cellular Automata Approach

Kunjam Nageswara Rao, Madugula Divya, M. Pallavi
and B. Naga Priyanka

Abstract The key feature of artificial life is the idea of emergence, where new patterns or behaviors emerge from complex computational processes that cannot be predicted. Emergence initiates the formation of higher-order properties via the interaction of lower-level properties. Biological networks contain many theory models of evolution. Similarities between the theoretically estimated networks and empirically modeled counterpart networks are considered as evidence of the theoretic and predictive biological evolution. However, the methods by which these theoretical models are parameterized and modeled might lead to inference validity questions. Opting for randomized parametric values is a probabilistic concern that a model produces. There persists a wide range of probable parameter values which allow a model to produce varying statistic results according to the parameters selected. While using the phenomenon of cellular automata, we tried to model life on a grid of squares. Each square in the grid is taken as a biological cell; we have framed rules such that the process of cell division and pattern formation in terms of biological theoretic perspective is studied. Relatively complex behaviors of the cell patterns which vary from generation to generation are visually analyzed. Three algorithms—game of life, Langton's ant, and hodgepodge—have been implemented whose technical implementation will provide an inspiration and foundation to build simulators that exhibit characteristics and behaviors of biological systems of reproduction.

K.N. Rao (✉) · M. Divya · M. Pallavi · B. Naga Priyanka
Department of Computer Science and Systems Engineering,
Andhra University, Visakhapatnam, India
e-mail: kunjamnag@yahoo.com

M. Divya
e-mail: vanquisher1607@gmail.com

M. Pallavi
e-mail: pallavi.patty@gmail.com

B. Naga Priyanka
e-mail: priyankabetha0@gmail.com

© The Author(s) 2015   73
N.B. Muppalaneni and V.K. Gunjan (eds.), *Computational Intelligence
Techniques for Comparative Genomics*, Forensic and Medical Bioinformatics,
DOI 10.1007/978-981-287-338-5_6

Keywords Evolution · Natural selection · Artificial life · Modeling life · Artificial ecosystem · Cellular automata · Game of life · Langton's ant · Hodgepodge

1 Introduction

Computers today are used in all aspects of modern life. One of the major areas where computers are widely used is coding. Many philosophical and theoretical questions can be answered via modeling by the means of coding. The roots of evolutionary design spring from the interaction between computer science, design and evolutionary biology [1]. The broad areas of art, design, and architecture are widely being used to build various applications today. A few of the applications include transforming images, constructing architectural forms, constructing 3D images and sculptures, applying various compositions, and visualization. The process involves designing and applying rules based on real-time observations. The rules governing the phenomenon are used to generate a network which models the real-time scenario via a machine. Rules or the instructions are framed such that they are straight forward and unambiguous. Simple local rules can often be used to create unexpected and complex patterns. The central feature of framing the rules is to explore infinite number of variations existing within a complex system [1]. Fast iterations are explored via rules, and thereby, performance of the object is modeled. The object behavior is merged with the complex interactions of the environment while tracing and applying the applicable rules. Some behaviors which cannot be predicted in advance due to persisting variations existing within the rules are modeled with the variations, and the results are interpreted to further understand the behavior of the environment. Repetitive tasks are built using looping constructs of programming, to form digital patterns, [2] where the interaction between two or more entities in an environment determine the future patterns or object states. Digital patterns explore emergence which involves formation of higher-order properties via interaction of lower-level elements.

An idea is conceptualized and turned into code using rule-based art and algorithmic thinking [2], where a method or a procedure is described in an ambiguous manner. Initially, a tangible physical process is taken which is abstractly translated into a form which is suitable to express a process as an algorithm. The transition can be generated in real time by means of a machine [3]. The algorithm is further implemented as code using simple rules which are repeated over multiple objects often generating complex phenomena. While expressing the algorithm, the degree of simplification, proper judgment of important events, and simple patterns take care of randomness and variation which are expressed when an unpredictable event occurs. Decanting of a neurological pattern within a phenomenon being measured and organism mutations creating new strands of organisms are a few other random

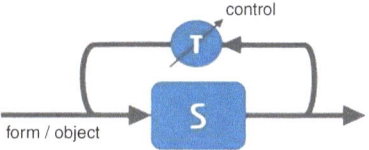

Fig. 1 Feedback system

events which disclose the measure of uncertainty persisting with in the system. Random events occurring in the biological world result in unexpected outputs.

During the ancient times, randomness was entangled with chance and fate. It is based on superstition. Random mutation occurring in the organisms of the biological world change the organism genetic structure, create variations, and produce unpredictable outcomes; the random walk algorithm [4] can be used to model all forms of cultural activities which include ecological, economical, biological, physical, and social activities; the central focus of the algorithm is systematic generation of mathematical equations to create variation. It uses the concepts of mathematics and physics to show object interactions. Randomness can be easily understood and visualized in real time using the constructs of this algorithm, thus giving the explorers an insight into the workings of a still unexplained phenomenon; real-time environment is visualized and coded as objects, which represents the entities of a program. Each object encapsulates the associated data and represents the interactions with other objects while exchanging messages within the environment. A class comprising of name, attributes, and methods encapsulates the states and behavior of the objects. Interactivity between abject is performed using message passing.

Ideas coming from life sciences and biological studies make use of creative coding to model the real-time environment which helps in understanding the phenomenon better. The coding constructs or frame rules are applied to generate visualization model using variations. These are modeled phenomenon and cannot be taken as the accurate stimulation of the phenomenon as no scientific experiments are being conducted. Feedback is a powerful process used in the construction of generation which offer rich possibilities of variations; that is, the organisms which grow, metabolize, reproduce, and respond to the activities within the environment is not a simulation. It is a new realm comprising of its own logic, life forms, and relational dynamics whose nature is entirely programmed and self-contained [5]. We begin with a system where, an object is fed into the system "s" and undergoes a transformation or variation "t". The generated results are further fed back into the same system. When the process is repeated over many iterations, the original form changes potentially, leading to increasing complexity or change. Acquiring a little degree of control over "t" also allows further variation to occur [2]; the programming of feedback is equivalent to recursion which is a neat conceptual programming trick (Fig. 1).

Fig. 2 Darwin's law of
natural selection

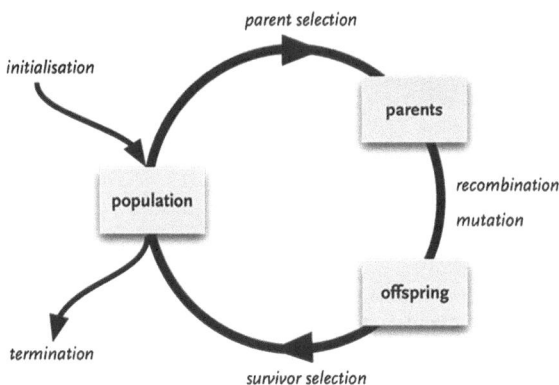

2 Related Work

Charles Darwin proposed the theory of evolution and introduced the concept of
natural selection where there exists a common ancestor for all the organisms and the
mutation caused while copying the genetic information leads to variations or randomness, which thereby generates a wide range of mutated organisms deriving their
genetic data and ancestral interface from their parents. Darwin proposed the law of
natural selection [1] where he claims that not all generated species survive to
produce further generation. Only the species with greater fitness ratio survive to
produce further offsprings and take part in evolution. According to Darwin's theory,
an algorithm of the array of life on earth has been proposed; this is called the
evolutionary algorithm [2] which primly frames the evolutionary esthetic designed
principles required to design creatures on a computer using computer programming
(Fig. 2).

   In the programmed model, many copies of the original randomly initialized cells
are made. Each copy may slightly differ with each other due to small changes in the
network connections and neural responses. Each object is tested individually, the
creatures that survive for longer periods become the parents for the next generation.
None of the other generations do well, but there will be some variation in the ability
to sustain due to random mutation occurring between the siblings. This process is
repeated over and over, and each generation gets better and better. Many different
variations of the regulatory algorithm have been developed, but the straightforward
algorithm comprising of the Darwin's law of natural selection has to obey three
conditions [2]:

1. A population of individuals that can make copies of themselves exists
2. The coping process is imperfect and may lead to variation
3. Variation and copying leads to differences in the ability of the offspring

   This process of evolution is not limited to biological life; almost anything can be
represented on a computer using the evolutionary algorithm. For the use of this

algorithm on a computer, the population of the individuals required is usually generated randomly or from a digital genome that controls the form, behavior, or function of an individual [2].

/ Procedure: Evolutionary Algorithm */*

Initialize the population with random individuals;
Evaluate the fitness of each individual;
Repeat until end conditions are satisfied;
{

  Select parents;

  Recombine pairs of parents;
  Mutate the resulting off spring;
  Evaluate fitness of the new individuals;
  Select individuals to survive to the next generation;

}

The fitness of an individual is the criteria of selection which is often determined on a problematic basis; fitter individuals are more likely to have more offsprings, but less fit individuals still have a chance of evolution; selective breeding and natural selection evaluate the fitness of the individuals by using different methods involving different algorithms which are framed and modeled with different rules and requirements. Artificial evolution and esthetic selections are the interactive generic algorithm that often takes the assistance of a human in the evolutionary loop because people are good at judging. Esthetics are what they are trying to evolve, whereas computers are not [2], but people are slow at evaluating fitness when compared to the computer which can efficiently compare thousands of individual image patterns very fast without getting tired. Sonic ecosystem is an artificial life artistic system which is modeled to capture the concepts of an open-ended biological system. This open-ended system is highly reactive to its environment and takes the help of esthetic selection of the user interacting with the system [6]. Eden is an example of interactive and self-generated artificial ecosystem which uses the model of cellular automata for music composition. Here, the entities populate a virtual world and compete for the existing limited resources. Part of their genetic representation permits the creatures to make and listen to sounds. Complex musical and sonic relationships develop as the creatures use sound to aid in their survival and mating process [7].

Using computer to model and produce simulated images and effects is an effective procedure of modeling where wide range of modeling tools are available to produce sculptures, 3D and 2D structures, and interaction between objects of an environment via computer space, governed by rules which are an outcome of our observations, and the model exists only in imagination. The physical and chemical dynamics of all kind of processes and the genetic evolutionary system of organic life continuously create new and original forms [3]. These dynamic forces in a biological world influence translation into real-world structures. Complex patterns are generated using mathematical formulae. All the imaginary forms can be designed on a computer. Computer keyboard and monitor are used as artistic and esthetic media [2]. The idea which is to be modeled is initially identified and is implemented as a program via the abstract rules framed at the time of design. The program when run gets converted into a stream of binary numbers indicating complex geometry which in turn gives the illusion of depth and space. The primary questions which are to be answered during the developmental scenario tend to be the following: firstly knowing how a designer exactly creates a virtual artificial life that evolves toward some subjective criteria of audience experiencing it without them needing to explicitly perform fitness selection, and secondly how the relationship between real and virtual spaces is acknowledged in a way that the spaces are integrated phenomenally.

The artificial life environment operates over a cellular lattice inhabited by agents who use an internal, evolvable, rule-based system to control the external environmental behavior. The use of genetic algorithms evolves toward finding maxima in fitness which is individually evaluated for each phenotype of the population [1]. Artificial life is a term first used by the computer scientist Chris Langton in 1986 to describe human-made systems that describe characteristics in biological systems. The domain is concerned with generating lifelike behavior using computer simulations or building robots [2]. According to Langton, living system exists independently and has to be redefined based on the instance of life as we know it or a more general prediction of life as it could be. Life is here defined by the mechanisms and behaviors that might be realized in other media such as silicon. Although silicon-based life sounds profoundly immoral, it redefines what life is and what life could be. Emergence is the central concept of artificial life which explains the crucial leap it makes between life and non-life [8]. The idea of emergence began with simple local rules through local interaction and feedback. Feedback allows complexity to emerge spontaneously via recursive looping which magnifies deviations and derives complex interactions and unpredictable evolutions associated with emergence [9].

Cormack's sonic Ecosystem "Eden" [7] has already been quoted before in the paper. Dawkins proposed the biomorph model [10] where he used a recursive tree drawing program to draw a single vertical line. The line further splits into two subbranches, and the process continues iteratively from generation to generation. All the sub-branches are mutations of the parent branches. The degree of mutation depends on the numeric value that controls the length and the direction of subbranches. Sims [11] used mathematics equations to generate images and said that artificial evolution has the potential to achieve flexible complexity. He explored the difficulty in automatically exploring esthetic success of a genetic image. His success allowed the user to guard the evolution of an image in a particular direction. Image interfacing allowed random selection of the two parent images which were used to breed a new population of offspring with inherited genetic traits. A system was developed by Todd and Latham [12] to evolve 3D forms through geometric procedures. The system allowed Latham to test and exhaust a formal grammar of any particular 3D form or structure (Fig. 3).

Fig. 3 Eden, Biomorph model, Karl Sims genetic image, William Latham's 3D form

2.1 Algorithms to Visualize Artificial Life

Cellular automata is a discrete model used in the field of computer science to model von Neumann (self-reproducing) comprising of a grid of squares where each square is termed to be a cell. Each cell has a state which is interpreted using rules which relatively evolve the cell to illustrate complex behavior and form structures. The set of neighboring cells, states, color pallets, and the rules defined to model the environment play a major role in effecting the behavior of the automaton. It makes a great initial model to build a system of many objects with varying states over time. Most important detail of how cellular automata works relates to time. Time does not refer to real-world time, but it is about cells on a grid living over a period of time called as the "generation." The cell's new state is a function of the entire cell's neighborhood at the previous generation which mathematically is represented as [13]

$$\text{CELL state at time } t = f(\text{CELL neighborhood at time } t - 1).$$

The edge cells of a grid remain constant, wrap around, or have different neighborhoods with different rule sets all together. Different rule sets produce different and complex patterns which vary from generation to generation [13]. Excluding traditional cellular automata, various other variations can also be used for modeling. Non-rectangular grids can be used, cellular automata can be probabilistic in nature where exact outcome need not be defined previously, and it can be continuous where the order of state values can range from 0 to 1, allowing the use of floating-point values for the state as used in hodgepodge algorithm [14–16]. Image processing algorithms operate on cellular automata like rules where operation performed on a pixel has direct effect on the neighboring pixels. Cellular automata can also relate to idea of complex adaptive system and be historic in nature where the current and the previous states of a cell are to be tracked. Cells rather than having a fixed position on a grid can be implemented as movable such that cells move about the screen freely without sticking to a fixed position. Systems can also be made nested to make modeling effective; for example, organ is a complex system of cells [13].

2.1.1 Game of Life Algorithm

John Conway created a system which today is known as the "Game of life"
[17, 18]. He explored possibilities of constructing a system which is capable of
performing universal computations. He constructed a rectangular grid comprising
of squares which he named as cells. He assigned vales 0, 1 to each square cell and
termed cells with value 0 dead cells and value 1 alive. The other probable possibilities of differing states include all cells becoming dead, system coming into a
steady state and becoming fixed thereafter, or system entering into oscillating states
which repeats itself for some periods of time. He programmed the system such that
it regularly updated its cells basing on a set of rules which he named as "life and
death rules." He stated that a cell that is alive can remain alive only if it has exactly
2 or 3 neighboring cells that are alive. A dead cell can become alive if it has exactly
3 neighboring alive cells. All other cells remain dead [17].

```
/* Procedure: Game of Life Algorithm */

Construct the Square Lattice;
Create and Initialize the initial position;
Repeat until end conditions are satisfied;
{
   For each cell in the lattice
    {
       var =0;
       For each neighboring cell
         {
             If current cell state =
             alive var++;
         }

       If the cell is in alive state
         {
       If ((var !=2) || (var
         !=3)) Kill cell;
       }

   Else /* the cell is dead */
   {
       If var =3
       Revive cell
   }
 }
}
```

He framed the rules trying to model the biological replication of the cells and
also stated that any live cells with fewer than 2 neighbors die as if caused due to
underpopulation. Live cells which have 2 or 3 live neighbors live onto the next
generation. If a cell has more than 3 neighbors, it dies due to overcrowding. A dead

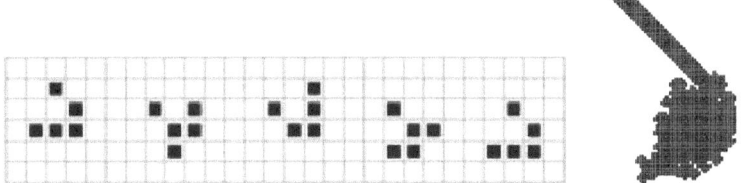

Fig. 4 Game of life [17, 18] pattern glider, Langton's ant trajectory [19] after 1,200 generations

cell with exact 3 neighbors becomes alive as if by reproduction. Initial pattern which is randomized is the seed of the system's iterative working [18] (Fig. 4).

2.1.2 Generalized Langton's Ant Algorithm

Chris Langton proposed another model of an artificial life which he named as Langton's ant [19–21]. He opted ant to be an agent who moves by turning left or right according to the color of the square it is heading onto. The color of the square changes accordingly. He took a square lattice and defined each square in the lattice as a cell similar to the game of life algorithm. He also defined that each cell has two states "to-left" and "to-right" which is determined according to the direction in which the ant moves. He framed the rules such that if an ant reaches a white square, it turns right, flips the color of square to black, and moves forward. If the ant reaches a black square, it turns left, flips the color of square to white, and moves forward. In the first 100 generations of the algorithm, rather simple and symmetric patterns are produced. Further patterns get complicated, and irregular patterns are produced. The ant traces a pseudorandom path until 10,000 steps. Finally, recurrent highway patterns are produced which repeat indefinitely.

```
/* Procedure: Generalized Langton's Ant Algorithm */

Construct the Square Lattice;
Create and Initialize the position of the Ant;
Teach Ant How to Behave;
Repeat until end conditions are satisfied;
{ If Square color = Black
   {
     Turn 90° Right
     Set Square Color =White
   }
   Else
   {
     Turn 90° Left
   Set Square Color =White
   }
}
```

2.1.3 Hodgepodge Algorithm

Hodgepodge algorithm [14–16] uses excitable cellular automation to model physical and chemical systems like oscillations and chemical reactions. Techniques like neural networks and wave dynamics can be used to increase the efficiency of the phenomenon. In excitable cellular automaton, study systems with uniform states which are linearly stable but are susceptible to finite perturbations exist. The cells within the lattice exist in three states—exited state (infected cells), recovering state (ill cells), and relaxed state (healthy cells). Each cell keeps fluctuating in between the three states on the basis of the states of its neighboring cells. Here, a $S \times S$ square lattice is used with periodic boundary conditions. Mathematically, each cell is assigned with a value which may range from 0 to $k - 1$. A healthy cell is assigned with 0, ill cell is assigned with $k - 1$, and infected cell can be assigned with any nonzero value lying in between 0 and $k - 1$. The basic rules applicable for this algorithm include the remodeling of biological rules. An ill cell eventually becomes healthy. A healthy cell can get infected or might remain healthy. An already infected cell can become more infected. The change in state is calculated basing on the fitness of the cell and its neighborhood.

/ Procedure: Hodge Podge Machine Algorithm */*

Initialize the population with random individuals;
Randomize the grid if it were the first generation;
Evaluate the fitness of each individual;
Iterate recursively to form next generations;
{

  For each cell in the grid get the infection level of the
  cell; Get the state of the cell
  {
  If Cell = fully ill, make it better;
  If Cell = partially ill, make it more sick;

  If cell = healthy, see if the neighboring infected cells are enough to make it
  ill; } / evaluate fitness iteratively and determine state */*

  Construct the cellular grid with next generation;
}

For all the three algorithms stated above, the main design criteria is to choose the rules carefully such that there is no explosive growth of the cells and there exist small initial patterns with unpredictable results which shows us a visual plot of the biological cell division ranging through generations. Simple rules are to be framed which adhere to the constraints of the biological scenario which is being tried to be modeled.

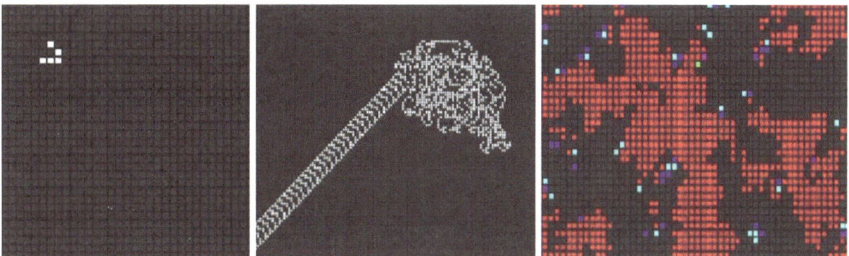

Fig. 5 Game of life, Langton's ant, and hodgepodge model visual grid simulations

3 Results

A system which attempts to integrate the open-ended nature of synthetic evolutionary systems into a virtual environmental place has been produced. The approach used in this paper has been to measure the components of the real environment and incorporate them into the virtual space, thereby opening the doors to enable and study evolutionary relationships between virtual agents. Simulation of laws of physical world is substituted by defining an artificial nature with fictional laws constituting a complete world of its own. This has been done using three approaches, results of which have been shown. In the game of life algorithm, little blobs move across the screen, sometimes dying, sometimes multiplying.

In Langton's ant algorithm, an ant makes frieze-like patterns and initially behaves to have well-known idea, but about the direction of movement in the later generations, it seems to have lost in an unknown city and keeps turning around. Non-equilibrium thermodynamics is demonstrated through hodgepodge algorithm by means of colored patterns. It is used to visualize results of reactions and oscillations of various kinds of biological material (Fig. 5).

4 Conclusion

By the process of evolution, we are exploring a multidimensional space of millions of sculptures with variations of varieties, all beyond the scope of our imagination. Computer is used for the development of artificial worlds with self-organizing properties. The simulative results help evaluating the theoretic biological properties visually. They act as an evidence for the theory models which help in enhancing the predictive action of biological evolution by means of comparisons.

5 Future Work

Esthetics is a challenging research problem which has a broader scope to be worked on; while modeling biological systems, many complex relationships exists between the organisms beyond mere competitive behavior between organisms. Cooperation, symbiosis, parasitism, and codependence can be behaviorally modeled. Many other relationships between the organism and the environment can be identified for modeling, hence deriving opportunities for research and development in a wider perspective. We today have challenged to create vivid artificial worlds in allowance with technology and to design them in a refined manner such that they awaken new possibilities of experiencing nature.

Acknowledgments We would like to express profound gratitude to Sri Kunjam Nageswara Rao, for his guidance, supervision, and generosity all through the study. We pay equal debt of gratitude to Professor P. Srinivasa Rao, Head of the Department, for providing invariable support and facilities. We are greatly thankful to the other faculty members of the department for their constant encouragement and valuable suggestions. We also thank S. Vakkalanka sir, Asst. Prof. Avanthi Institute of Engineering and Technology, for his suggestions while framing the paper.

References

1. Bentley PJ (ed) (1999) Evolutionary design by computers. Morgan Kaufmann, Los Altos
2. Web Link to Future Learn (Creative coding: Monash University). https://www.futurelearn.com/courses/creative-coding
3. Driessens E, Verstappen M Ima traveller, website http://notnot.home.xs4all.nl/
4. Web Link to Online Visualization insights. https://www.random-walk.com
5. Haru JI, Graham W (2014) Artificial nature, 14 Aug 2014. http://artificialnature.mat.ucsb.edu
6. Cormack JM (2003) Evolving sonic ecosystems. In: Adamatzky A (ed) The international journal of systems and cybernetics—kybernetes, vol 32 no. 1/2. Emerald, Northampton
7. Cormack JM (2001) Eden: an evolutionary sonic ecosystem. In: Kelemen J, Sosik P (eds) ECAL 2001. LNCS, vol 2159, pp 133–142. Springer, Heidelberg
8. Whitelaw M (2004) Metacreation: art and artificial life. MIT Press, Cambridge
9. Hayles NK (1999) How to become Posthuman: virtual bodies in cybernetics, literature and informatics. University of Chicago Press, Chicago
10. Dawkins R (1996) The blind watchmaker. W.W. Norton & Company Inc., New York
11. Sims K (1991) Artificial evolution for computer graphics. In: Proceedings of SIGGRAPH91 computer graphics annual conference series. ACM SIGGRAPH, Las Vegas, New York
12. Todd S, Latham W (1999) The mutation and growth of art by computers. In: Bentley PJ (ed) Evolutionary design by computers. Morgan Kaufmann, Los Altos, pp 221–250
13. Daniel S Web link to nature of code. www.natureofcode.com
14. Rafael PS, Winfer CT, William ESY (2002) Parallel implementations of cellular automata algorithms on the AGILA high performance computing systems. In: Proceedings of the sixth international symposium on parallel architectures, algorithms, and networks (I-SPAN'02). IEEE Computer Society Press, USA, pp 125–131
15. Reiter C (2009) With J: the Hodge Podge machine, vol 130(41)
16. Jaime S, Eduardo RM Algorithmic sound composition using coupled cellular automata. Interdisciplinary Center for Computer Music Research (ICCMR), University of Plymouth, UK
17. Adamatzky A (ed) (2010) Game of life cellular automata. Springer, London

18. Lichtenegger K (2005) Stochastic cellular automaton models in disease spreading and ecology
19. Jean PB (2001) How fast does Langton's ant move? J Stat Phys 102:355–360
20. Tatsuie T, Takeo H (2011) Recognizing repeatable configurations of time reversible generalized Langtons Ant is PSPACE—Hard. Algorithms 4(1):1–15
21. Gajardo A, Moreira A, Goles E Complexity of Langton's Ant. Discrete Appl Math 117 (1):41–50

Identification of Deleterious SNPs in TACR1 Gene Using Genetic Algorithm

Dharmaiah Devarapalli, Ch. Anusha and Panigrahi Srikanth

Abstract Bioinformatics is a specific research and development area. The purpose of bioinformatics mainly deals with data mining and the relationships and patterns in large databases to provide useful information analysis and diagnosis. Single nucleotide polymorphisms (SNP) are one of the major causes of genetic diseases. Identification of disease-causing SNPs can identify better disease diagnosis. Hence, the present study aims at the identification of deleterious SNPs in TACR1 gene. Developing an algorithm plays a vital role in computational intelligence techniques. In this paper, a genetic algorithm (GA) approach is to develop rules and it is presented. The importance of the accuracy, sensitivity, specificity, and comprehensibility of the rules is simplified for the implementation of a GA. The outline of encoding and genetic operators and fitness function of GA are discussed. GA is using to identify deleterious or damaged SNPs.

Keywords SNPs in TACR1 · Computational intelligence technique · Fitness function · Genetic algorithm

D. Devarapalli (✉) · Ch. Anusha
Department of Computer Science and Engineering, Vignan's Institute of Information Technology, Duvvada, Viskhapatnam, Andhra Pradesh 530049, India
e-mail: deverapalli.dharma@gmail.com

Ch. Anusha
e-mail: anusha.ch045@gmail.com

P. Srikanth
Department of Computer Science and Engineering (Software Engineering), Vignan's Institute of Information Technology, Duvvada, Viskhapatnam, Andhra Pradesh 530049, India
e-mail: srikanth.panigrahi@gmail.com

© The Author(s) 2015
N.B. Muppalaneni and V.K. Gunjan (eds.), *Computational Intelligence Techniques for Comparative Genomics*, Forensic and Medical Bioinformatics, DOI 10.1007/978-981-287-338-5_7

1 Introduction

The genetic algorithm (GA) was first introduced by Holland and his team in the university of Michigan in the early 1960s [1]. It is widely used to solve many optimization problems in all fields of engineering and sciences. These algorithms are hypothetically and empirically proved to be giving efficient search in complex space. GAs are very useful, particularly when the size of the dataset is very large. Knowledge engineering can be used to solve a diversity of tasks like clustering, classification, and regression and association discovery [2]. Association rule mining is one of the widely used approaches for constructing the computational intelligence techniques. A classification rule is represented in the form if P then Q, where P is a combination of predicting attribute values and Q is the predicted class.

Basheer et al. [3] proposed a GA-based approach for mining classification rules from large database is presented for emphasizing accuracy, coverage, and comprehensibility of the rules and simplifying the implementation of a GA. The design of encoding, genetic operators, and fitness function of GA for this task are discussed. Experimental results show that GA proposed in this paper is suitable for classification rule mining and those rules discovered by the algorithm have higher classification performance to unknown data.

Pradhan et al. [4] proposed a multiclass genetic programming (GP)-based classifier design that will help the medical practitioner to confirm his/her diagnosis toward prediabetic, diabetic, and non-diabetic patients.

Permann [5], this paper tell us about initial research using GAs his help to optimize infrastructure protection and restoration decisions. This research suggests that it applies GAs for problem infrastructure modeling and observes in order to determine the optimum assets to restore other disaster. First, the problem space is introduced. Next, the change based on simulation used by the GAs is introduced. Then, the critical subnetwork concept, in GAs, and similar research are described. Finally, the GA for decision-making research is discussed.

Neurokinin 1 (NK1) is within a receptor for substance P. Substance P involved in pain transmission. NK1 is also known as TACR1 gene.

The neurokinin (NK) 1 receptor is a G protein-coupled receptor and member of the tachykinin family, which also includes NK2 and NK3 receptors. The NK1 receptor is localized in highly concentrated central nervous system (CNS) and particularly the striatum, and some hypothalamic and thalamic nuclei and peripheral tissues and uncureable renal, liver, and skin problems. NK1 receptors do not act directly to the central stress reactions, mood control, excitatory neurotransmission, immune modulation, and airway and lung function. The receptor displays greater potency for the endogenous agonist substance P than for neurokinin A and neurokinin B (SP > NKA > NKB). The human gene encoding the NK1 receptor has been localized.

TACR1 (tachykinin receptor 1 or neurokinin receptor 1) is a coding gene. Diseases associated with TACR1 include telogen effluvium, liver, skin problems, and peripheral nervous problems; these are related to super pathways: GPCR ligand binding and peptide GPCRs. GA annotations related to this gene include tachykinin receptor activity and substance P receptor (SPR) activity. An important paralogy of this gene is GPR83.

More than half of the origin human genetic diseases are caused by amino acid substitutions. So the present study is aimed at studying single nucleotide polymorphisms (SNP) in TACR1 gene that tells about of the gene and function of the protein. Defects in TACR1 gene can cause telogen effluvium, liver, skin diseases, and nervous pains. It is to identify SNPs which are deleterious for functioning of TACR1 protein and causing above-specified diseases [6–8].

2 Dataset Description

A SNP is a DNA sequence which is in the form of a A, T, C, and G in the genome (or other shared sequence), different members of a biological species, or paired chromosomes.

SNP are within coding sequences of gene, noncoding regions of gene, or intergenic regions (regions between genes). SNPs within a coding sequence do not change the amino acid sequence of the protein that is produced, due to loss of the origin genetic code.

SNPs in the coding region are of two types, synonymous and non-synonymous SNPs. Synonymous SNPs do not affect the protein sequence, while non-synonymous SNPs change the amino acid sequence of protein. The non-synonymous SNPs are of two types: (a) missense and (b) nonsense.

SNPs that are not available in coding regions also may still affect gene splicing, transcription factor binding, messenger RNA (mRNA) degradation, or the sequence of noncoding RNA. Gene expression affects the type of SNP is referred has an expression SNP (eSNP) and may be upstream or downstream from the gene.

NK1 is the recent study of medications that possesses unique antidepressant, anxiolytic, and antiemetic properties. The discovery of NK1 receptor antagonists was a turning point in the prevention of the vomiting. The registered clinical use of NK1 receptor antagonists was the treatment of emesis, associated with liver and skin diseases.

The tachykinin receptor 1 (TACR1) also known as NK1 receptor (NK1R) or SPR is a G protein-coupled receptor (GPCR) found in the CNS and peripheral nervous system (PNS). The endogenous ligand for this receptor is substance P, although it has some affinity for other tachykinins. The protein is the product of the TACR1 gene.

There are SNPs present in TACR1 gene (dbSNP database) [9]. There are in total 3,638 SNPs present in TACR1 genes. In this paper, we consider 20 of them which are coding non-synonymous SNPs. SNPs are present in 3′ UTR region and 5′ UTR. (SIFT tools using considered these followed to 2.17 % of total SNPs constitute non coding synonymous SNPs. 2.33 % SNPs are present in 3′ UTR region and 1.12 % in 5′ UTR region.)

Most of nsSNPs are missense mutations. There are two deleterious stop gain mutations at positions (Y [Tyr] Ter[*] [OCH]) and (Y [Tyr] Ter[*] [OCH]) (Figs. 1, 2 and 3).

SNP using TACR1 receptor following considered Dataset V → M, Valine is mutated to Methionine at position 253, Aspartic acid is replaced by Tyrosine in SNP D → Y, Alanine is replaced by Threonine A → T, Serine is replaced by Leucine S → L, Tyrosine is replaced by Cysteine Y → C, Proline is mutated to Tyrosine P → L. These are Missense, cds-synon, stopgain (deleterious).

Most of nsSNPs are missense mutations. There are two deleterious stop gain mutations at positions (Y [Tyr] Ter[*] [OCH]) and (Y [Tyr] Ter[*] [OCH]).

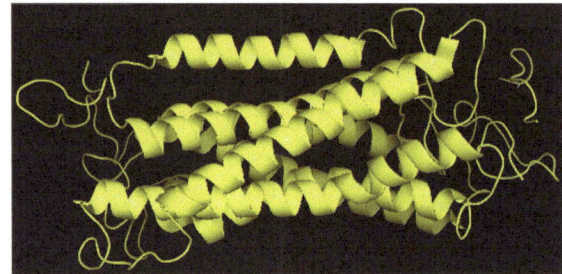

Fig. 1 TACR1 protein structure obtained from PDB (structure ID: 2KS9)

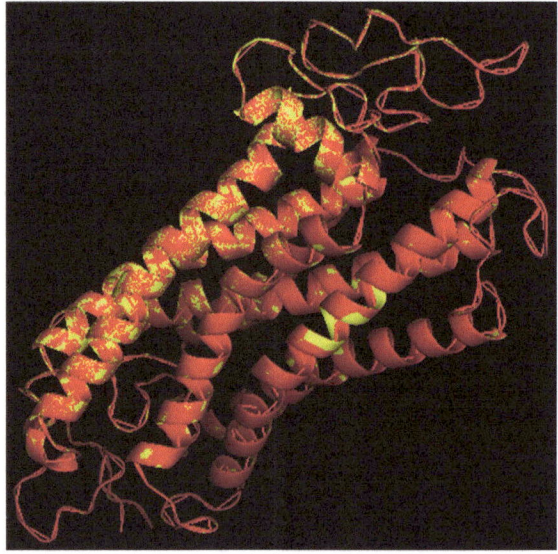

Fig. 2 Superimposition of native TACR1 protein (*red*) on V253 M mutant (*yellow*)

Fig. 3 Superimposition of native TACR1 protein (*red*) on D129Y mutant (*green*)

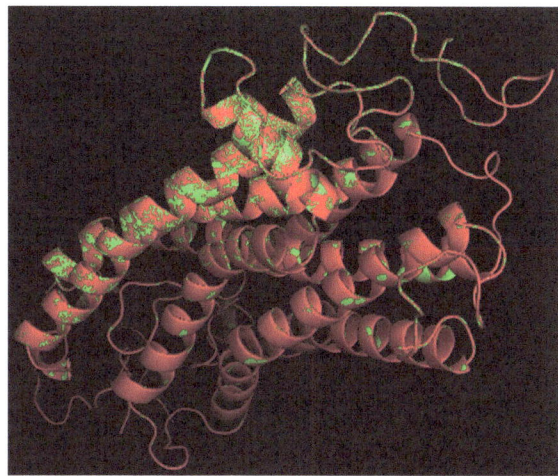

3 Methodology

3.1 Genetic Algorithm

GA proposed by J.H. Holland is one of the efficient optimization techniques which is based on 'Survival of the fittest.' In this paper, we used GA to generate individual classification rules from the given set of training data items. The individual rule consists of 'if part' and 'then part,' where 'if part' is known as antecedent and 'then part' is known as consequent [3, 10]. The rule representation is as follows.

If P then Q

where
P is the list of attributes involved in the rule formation.
Q is the class label of the rule.

Gene Representation:
In this approach, each individual is represented with conjunction of conditions composing a given rule antecedent. Each gene represents a rule condition of the form as follows.

$$A_i \text{ Op}_i V_{ij}$$

where
A_i represents the ith predictor attribute,
Op_i comparison operator $\{<, >, \leq, \geq, =\}$, and
V_{ij} denotes jth value of the ith attribute (Figs. 4, 5 and 6).

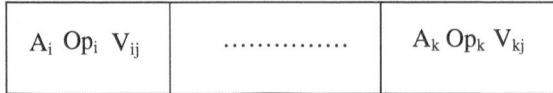

A_i Op_i V_{ij}	. . . .	A_k Op_k V_{kj}

Fig. 4 Individual genome representation

A	A	A	Classes (Two) Missense or Non Missense
1	2	3	

Fig. 5 Rule consisting of attributes

	#	

Fig. 6 Operator combination for the above attribute values

The absence of the attribute in the genome is represented with '#' symbol which indicates the absence of the value in the rule. A central mechanism in the GA is the fitness function that plays vital role in optimizing a specific problem.

Classification rule is in the form of P to Q

$P \wedge Q$ value is the number of samples in the dataset that are satisfying antecedent and consequent in the rule.

P value is the number of samples in the dataset that are satisfying only the antecedent part in the rule.

Q value is the number of samples in the dataset that are satisfying only the antecedent part in the rule.

Rule length calculation represents the number of attributes involved in rule formation.

L is the maximum possible length of the rule (no. of attributes).

B is the length of the current rule.

True Positives (TP): This refers to the positive tuples that were correctly labeled by the classifier.

True Negatives (TN): This refers to the negative tuples that were correctly labeled by the classifier.

False Positives (FP): This refers to the negative tuples that were incorrectly labeled as positive.

False Negatives (FN): This refers to the positive tuples that were mislabeled as negative.

Sensitivity: It measures the ability of the method to identify the occurrence of target class accurately.

Specificity: It measures the ability of the method to separate the target class.

Comprehensibility: The comprehensibility of the rule R can be defined by the number of attributes on the left-hand side of the rule.

$$TP = P \wedge Q; \quad FP = P - (P \wedge Q); \quad FN = Q - (P \wedge Q).$$

$$TN = (\text{Number of samples in the data set}) - P - Q + P \wedge Q.$$

$$\text{Sensitivity} = \frac{TP}{TP + TN};$$

$$\text{Sensitivity} = \frac{TN}{TN + FP};$$

$$\text{Comprehensibility} = \frac{L - B}{L - 1}$$

Fitness = Sensitivity * Specificity + 0.2 * (Comprehensibility).

3.1.1 Population

Select the new individuals from the new generation. According to this method, the individuals in a population are randomly shuffled, and then, the crossover operation is applied on each mating pair which produces two new offspring. The parent and the offspring with the highest fitness value are selected to be mutated with a probability. In our work, we used tournament selection technique for selecting the offspring individuals.

3.1.2 Tournament Selection

In GA, the common and robust selection mechanism used to select the individuals is tournament selection approach. This technique is used to predict the convergence rate of gap utilizing tournament selection.

This technique is used to predict the convergence rate of gap utilizing tournament selection.

Tournament selection steps:

(i) Select any two rules (pool-1) randomly from the first half and second half (pool-2) of the initially generated rules.
(ii) Select the parents with highest fitness value from pool-1 and pool-2, and form them as pool-3.
(iii) Apply crossover process on pool-3 rules.

3.1.3 Crossover Operation

The crossover operation takes two individuals as input expressions and selects a random point and interactions the subexpression at the rear point. While implementing the crossover operation, there may be chance of having errors like a crossover may produce individuals in which an attribute may be involved more than once and the offspring may be the same as that of the parent expression with a probability of 1. To overcome these errors, we used the below-mentioned techniques while performing crossover operation.

3.1.4 Mutation Operation

After crossover mutation takes place, it is the genetic operator that introduces genetic diversity in the population.

Mutation takes place whenever the population tends to become homogeneous due to repeated use of reproduction and crossover operator. This can produce the entirely new gene values being added to the gene pool. With this new gene values, the GA may be able to produce better solution than the previous one.

Steps from mutation with attribute modification:

(i)   Select a random point within the attribute range.
(ii) Form the new rule by changing the selected attribute value.
(iii) Compute the fitness of the newly formed rule; if the fitness is greater than the parent, then add this rule to the initial population.
(iv) Repeat the above process for required number of times.

Sequential steps in mutation with attribute removal:

(i)   Select a random point within the attribute range.
(ii) Form the new rule by removing the attribute value at the selected point.
(iii) Compute the fitness of the new rule. If the fitness of the new rule is greater than the parent, then add this chromosome to the initial population.
(iv) Repeat this process for required number of times.

3.2 Stopping Criteria

- If all the records in the dataset belong to the same class
- If dataset has all similar records in the attribute list
- If the dataset is empty.

Table 1 Performance metrics

Measure	Formula
Accuracy	$\dfrac{TP + TN}{TP + TN + FP + FN}$
Sensitivity	$\dfrac{TP}{TP + FN}$
Specificity	$\dfrac{TN}{TN + FP}$

3.3 Performance Metrics

Most widely used performance measures in the computational intelligence techniques are accuracy, sensitivity, and specificity. The ability of the model to correctly predict the class label of previously unseen or new data is defined as accuracy. Accuracy, sensitivity, and specificity are measures as follows (Table 1).

4 Results

The performances of the GA with the traditional classification methods we calculate of the accuracy. GA approach to construct computational intelligence techniques for the diagnosis of Deleterious. We identified major mutations in TACR1 gene—rs17838409—V253 M, rs77755890—D129Y, rs111831416—Y92C.

5 Conclusion

The proposed GA of fitness function developed using good methodology is found to be effective in diagnosis of clinical condition. SNPs in TACR1 data is considered Experimental validation of the proposed system. The rule based on Determines the SNPs in TACR1 Gene condition based on the Missense or Nonsense. GA using developed techniques has been used for identification of deleterious SNPs in TACR1 gene.

References

1. Holland JH (1975) Adaptation in natural and artificial systems. University of Michigan Press, Ann Arbor (s.l)
2. Han J, Kamber M, Pei J (2011) Data mining: concepts and techniques. Morgan Kaufmann, Los Altos (Third s.l)
3. Al-Maqaleh BM, Shahbazkia H (2012) A genetic algorithm for discovering classification rules in data mining. Int J Comput Appl 41(18):40–44 (0975–8887)

4. Pradhan MA, Bamnote GR, Tribhuvan V, Jadhav K, Chabukswar V, Dhobale V (2012) A genetic programming approach for detection of diabetes. Int J Comput Eng Res (ijceronline.com) 2(6):91
5. Permann MR (2007) Genetic algorithms for agent-based infrastructure interdependency modeling and analysis. INL/CON-07-12317, SpringSim
6. Datar P, Srivastava S, Coutinho E, Govil G (2004) Substance P: structure, function, and therapeutics. Curr Top Med Chem 4(1):75.103. doi:10.2174/1568026043451636 (PMID14754378)
7. Ng CP, Henikoff S (2001) Predicting deleterious amino acid substitutions. Genome Res 11:863–874
8. Ng CP, Henikoff S (2003) SIFT: predicting amino acid changes that affect protein function. Nucleic Acids Res 31:3812–3814
9. http://www.ncbi.nlm.nih.gov/SNP/
10. Carvalho DR, Freitas AA An immunological algorithm for discovering small-disjunct rules in data mining

Author Biographies

Dr. Devarapalli Dharmaiah is currently working as an Associate Professor in Computer Science and Engineering department, Vignan's Institute of Information Technology, Visakhapatnam and has the teaching and research experience of about 10 years. He also guided various dissertation works for both UG and PG students of VIGNAN'S IIT and other Colleges. He has taught various subjects of Computer Science and Applications for both UG & PG students such as C programming, Data Structures, Java, Operating Systems, Compiler Design, Linux, UNIX and Bioinformatics. He has a very good reputation among the students and faculty community for his proficiency in subjects. He is a lifetime member of CSI. He has published many papers in National, International conferences and leading Journals. His areas of interest are Bioinformatics, Neural Networks, Data Mining and Computer Networks.

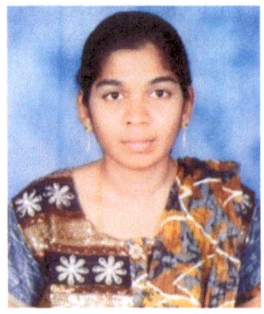

Miss. CH. Anusha is currently pursuing her 2 years of M.Tech (CSE) in Department of Computer Science and Engineering at Vignan's Institute of Information Technology, Visakhapatnam. Her current research is on Medical Diagnosis. Her areas of interest include Bioinformatics and Data Mining.

Mr. Panigrahi Srikanth   is currently pursuing his 2 years of M.Tech (SE) in Department of Computer Science and Engineering at Vignan's Institute of Information Technology, Visakhapatnam. He received his B.Tech (IT) from Gokul Institute of Technology and Sciences, Piridi, Bobbili, Vizianagaram, A.P. His current research is on Medical Diagnosis (Heart, Hepatitis, Diabetes, Thyroid and Cancer Diseases). His areas of interest include Bioinformatics, Data Mining and Information Security.

Identification of AIDS Disease Severity Using Genetic Algorithm

Dharmaiah Devarapalli and Panigrahi Srikanth

Abstract Bioinformatics is a data intentionally the field in Research and Development. The purpose of bioinformatics data mining (DM) is to observe the relationships and patterns in large databases to provide useful data analysis and results. Evolutionary algorithms play a main role in computational intelligence techniques. An developing situation was created throughout the world regarding the human immunodeficiency virus infection/acquired immunodeficiency syndrome (HIV/AIDS) disease that is mainly stigma. Every country is facing this problem. According to present survey is World health organization (WHO), AIDS disease has its complexity are health disease going present century. A best way to early examine of AIDS may improve the lives of all people affected by AIDS and people may lead healthy life. In this part, we have present an evolutionary algorithm known as Genetic Algorithm (GA) for better results of AIDS disease using association rule mining. In this computational intelligence technique, we tested the performance of the method using AIDS dataset. We presented a better fitness function using coverage, comprehensibility, and rule length. This fitness function we achieved is promising accuracy for model.

Keywords Artificial immune system · AIDS of CD4 cell count · Computational intelligence technique · Fitness function · Genetic algorithm

D. Devarapalli · P. Srikanth (✉)
Department of Computer Science and Engineering, Vignan's Institute
of Information Technology, Duvvada, Visakhapatnam 530049, Andhra Pradesh, India
e-mail: srikanth.panigrahi@gmail.com

D. Devarapalli
e-mail: deverapalli.dharma@gmail.com

© The Author(s) 2015
N.B. Muppalaneni and V.K. Gunjan (eds.), *Computational Intelligence
Techniques for Comparative Genomics*, Forensic and Medical Bioinformatics,
DOI 10.1007/978-981-287-338-5_8

1 Introduction

Nowadays, humans are suffering from many health problems. This century's most progressive disease is HIV/AIDS and its problems. HIV infects only humans; the deficiencies lie in the body's immune system, which normally protects the body against illness and is a virus. Viruses are tiny substances that enter the body's cells and cause illness. After the virus enters the body, there is a period of rapid viral replication, leading to a drastic change of virus in the peripheral blood. During primary infection, the level of HIV may reach several million virus particles per milliliter of blood. This response is accompanied by a marked drop in the number of circulating CD4 T cells. Controlling virus levels, which peak and then decline, as the CD4 T cell counts recover. A good CD8 T cell response has been linked to slower disease progression and a better prognosis, though it does not eliminate the virus.

Artificial Immune Systems (AIS) is a new field of study devoted to the development of computational models based on the behavior of the biological immune system, applied to several Engineering and Computer Science problems. Some of its applications include pattern recognition, fault and anomaly detection, data analysis, agent-based systems, scheduling, machine learning, control and autonomous navigation, search and optimization methods, artificial life, and information systems security [1]. AIS can be regarded as collection of algorithms which are abstracted from natural immune system such as HIS [2]. AISs have been applied to anomaly detections [3, 4].

The Genetic algorithm (GA) was first introduced by John Holland and his team in the University of Michigan in the early 1960s [5]. It is widely used to solve many optimization problems in all fields of engineering and sciences [6]. These algorithms are hypothetically and empirically proved to be giving efficient search in complex space. GAs are very useful particularly when the size of the dataset is very large. Knowledge engineering can be used to solve a diversity of tasks such as clustering, classification, and regression and association discovery [7]. Association rule mining is one of the widely used approaches for constructing the computational intelligence techniques. A classification rule is represented in the form: If P then Q, where P is a combination of predicting attribute values and Q is the predicted class.

Varma et al. [8] GA for better diagnosis of diabetes disease using association rule mining. In this computational intelligence technique, we tested the performance of the method using the Pima Indian Diabetes (PID) dataset taken from UCI machine learning repository.

Basheer et al. [9] a GA-based approach for mining classification rules from large database is presented and used for emphasizing on accuracy, coverage, and comprehensibility of the rules and simplifying the implementation of a GA. The design of encoding, genetic operators, and fitness function of GA for this task is discussed.

Pradhan et al. [10] a multi-class genetic programming (GP)-based classifier design will help the medical practitioner to confirm his/her diagnosis toward prediabetic, diabetic, and non-diabetic patients.

Permann [11] this paper describes initial research using GAs and helps to optimize infrastructure protection and restoration decisions. This research proposes to apply GAs to the problem of infrastructure modeling and analysis in order to determine the optimum assets to restore or protect from attack or other disaster.

2 Dataset Description

Genetic research indicates that HIV originated in west-central Africa during the late nineteenth or early twentieth century. HIV infection was identified in the early part of the decade. Since its discovery, AIDS has caused an estimated 36 million deaths (as of 2012).

The World Health Organization (WHO) first proposed a definition for AIDS. The WHO classification has been updated and expanded several times, with the most recent version. The WHO Disease Control and Prevention also creates a classification system that uses HIV infections based on CD4 count and clinical symptoms of the infection in three stages:

Stage 1: CD4 count above 500 cells/mm^3 No AIDS Condition
Stage 2: CD4 count between 500 and 200. AIDS defining condition. (In this condition to considered ART centers).
Stage 3: CD4 count ≤200 cells/mm^3 or AIDS defining conditions. (Near Death Stage).

Unknown: If insufficient information is available to make any of the above classifications for surveillance purposes, the AIDS diagnosis still stands even if, after treatment, the CD4+ T cell count rises to above 200 per mm^3 of blood or other AIDS-defining illnesses are cured (Table 1).

3 Methodology

3.1 Genetic Algorithms

GA is one of the efficient optimization techniques which is based on "Survival of the fittest." In this paper, we used GA to generate individual classification rules from the given set of training data items. The individual rule consists of "if part" and "then part." The "if part" is known as antecedent and "then part" is known as consequent [9, 13]. The rule representation is as follows:

If P then Q where
P is the list of attributes involved in the rule formation and
Q is the class label of the rule.

Table 1 CD4 cell count classification

Levels CD4+ count cells/mm^3	Levels CD4+ count cells/mm	Interpretation
>500	>0.500	Acute primary infection
		Recurrent vaginal candidiasis
		Persistent generalized lymphadenopathy
<500	<0.500	Pulmonary tuberculosis
		Pneumococcal pneumonia
		Herpes zoster
		Oropharyngeal candidiasis
		Oral hairy leucoplakia
		Extra-intestinal salmonellosis
		Kaposi's sarcoma
		HIV-associated idiopathic thrombocytopenic purpura
<200	<0.200	Mucocutaneous herpes simplex
		Cryptosporidium
		Microsporidium
		Esophageal candidiasis
		Miliary/extrapulmonary tuberculosis
		HIV-associated wasting
		Peripheral neuropathy
<100	<0.100	Cerebral toxoplasmosis
		Cryptococcal meningitis
		Non-Hodgkin lymphoma
		HIV-associated dementia
		Progressive multifocal leucoencephalopathy
<50	<0.050	CMV retinitis/gastrointestinal disease
		Primary CNS lymphoma
		Disseminated MAI

AIDS patient CD4 cell count categories are different types which are used to develop AIDS patient dataset. However in the AIDS patients' dataset there are 100 instances, we have used only 20 instances in this paper. The dataset used for evaluation is given in Table 2

Gene Representation:

In this approach, each individual is represented with conjunction of conditions composing a given rule antecedent. Each gene represents a rule condition of the form as follows:

$$A_i \, Op_i \, V_{ij}$$

Table 2 AIDS patient Dataset

S.No.	Age	Gender	Marital status	CD4 cell count (cells/mm^3)	AIDS
1	21	Male	Unmarried	450	Positive (AIDS)
2	18	Male	Unmarried	650	Negative (No AIDS)
3	16	Female	Unmarried	350	Positive
4	35	Female	Married	202	Positive
5	25	Female	Married	400	Positive
6	22	Male	Married	150	Positive
7	29	Male	Married	100	Positive
8	10	Male	Unmarried	275	Positive
9	15	Female	Unmarried	350	Positive
10	65	Female	Married	65	Positive
11	50	Male	Married	120	Positive
12	45	Female	Married	185	Positive
13	12	Male	Unmarried	405	Positive
14	08	Female	Unmarried	375	Positive
15	27	Male	Married	750	Negative
16	23	Female	Married	850	Negative
17	32	Male	Married	750	Negative
18	19	Male	Unmarried	425	Positive
19	20	Female	Unmarried	550	Positive
20	55	Male	Married	55	Positive

AIDS patient dataset description is described using the following attributes: Age (children to old age persons), Gender (Male and Female), Marital Status (Married and Unmarried), and CD4 Cell Count is given in Table 1 and classes are AIDS is divided two types classes used on positive (AIDS) and Negative (No AIDS). Those are following developed [12]

where A_i represents the i-th predictor attribute,

Op_i comparison operator $\{<, >, \leq, \geq, =\}$ and V_{ij} denote j-th value of the i-th attribute (See Figs. 1, 2, and 3).

The absence of the attribute in the genome is represented with "#" symbol which indicates the absence of the value in the rule. A central mechanism in the GA is the fitness function that plays vital role in optimizing a specific problem.

Classification rule is in the form of: P to Q

$P \wedge Q$ Value is the number of samples in the dataset that are satisfying antecedent and consequent in the rule.

$A_i \ Op_i \ V_{ij}$	....	$A_k \ Op_k \ V_{kj}$

Fig. 1 Individual genome representation

A	A	A	A	Classes(Two)
1	2	3	4	Positive or negative

Fig. 2 Rule consisting of attributes

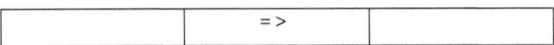
	=>	

Fig. 3 Operator combination for the above attribute values

P Value is the number of samples in the dataset that are satisfying only the antecedent part in the rule.

Q Value is the number of samples in the dataset that are satisfying only the antecedent part in the rule.

Rule length calculation represents the number of attributes involved in rule formation.

L is the maximum possible length of the rule (Number of attributes).

B is the length of the current rule.

True Positives (TP): This refers to the positive tuples that were correctly labeled by the classifier.

True Negatives (TN): This refers to the negative tuples that were correctly labeled by the classifier.

False Positives (FP): This refers to the negative tuples that were incorrectly labeled as positive.

False Negatives (FN): This refers to the positive tuples that were mislabeled as negative.

Sensitivity: It measures the ability of the method to identify the occurrence of target class accurately.

Specificity: It measures the ability of the method to separate the target class.

Comprehensibility: The comprehensibility of the rule R can be defined by the number of attributes on the left-hand side of the rule.

$TP = P \wedge Q$; $FP = P - (P \wedge Q)$; $FN = Q - (P \wedge Q)$.

$TN = $ (Number of samples in the dataset) $ - P - Q + P \wedge Q$.

$$\text{Sensitivity} = \frac{TP}{TP + TN};$$

$$\text{Specificity} = \frac{TN}{TN + FP};$$

$$\text{Comprehensibility} = \frac{L - B}{L - 1};$$

Fitness = Sensitivity * Specificity + 0.2 * (Comprehensibility) [8].

The AIDS parameters like in CD4 cell count are Marital status, Age, Gender, etc. By using these we developed several rules. For CD4 cell count parameter using followed Healthy condition, Pre-AIDS conditions are Medium, Low conditions, and finally AIDS condition (Near Death Stage). Those conditions are discussed below.

Rule 1: No AIDS Condition (Healthy condition)
If (CD4 cell count is above 500) and (Gender is Male or Female) and (Marital Status is Married or Unmarried) and (Age is between 8 and 80) then NO AIDS (Healthy condition).

Rule 2: Pre-AIDS Condition (Low condition)
If (CD4 cell count is between 500 and 350) and (Gender is Male or Female) and (Marital Status is Married or Unmarried) and (Age is between 8 and 80) then Low-AIDS (Pre-AIDS (Low) condition).
    Message for AIDS patients is considered to ART Centers.

Rule 3: Pre-AIDS Condition (Medium condition)
If (CD4 cell count is between 350 and 200) and (Gender is Male or Female) and (Marital Status is Married or Unmarried) and (Age is between 8 and 80) then Medium-AIDS (Pre-AIDS (Medium) condition).

Rule 4: AIDS Condition (Near Death Stage)
If (CD4 cell count is below 200) and (Gender is Male or Female) and (Marital Status is Married or Unmarried) and (Age is between 8 and 80) then AIDS (Near Death stage).

The "GA" process flow chart is shown in Fig. 4.

A. *Generation of Population*
    Elitist recombination technique is used to select the new individuals from the new generation. According to this method, the individuals in a population are randomly shuffled and then the crossover operation is applied on each mating pair which produces two new offspring. The parent and the offspring with the highest fitness value are selected to be mutated with a probability. In our work, we used tournament selection technique for selecting the offspring individuals.
B. *Selection*
    In GA, the common and robust selection mechanism used to select the individuals is tournament selection approach. This technique is used to predict the convergence rate of gap utilizing tournament selection.
    This technique is used to predict the convergence rate of gap utilizing tournament selection.
    Selection steps:

    i. Select any two rules (pool-1) randomly from the first half and second half (pool-2) of the initially generated rules.
    ii. Select the parents with highest fitness value from pool-1 and pool-2 and form them as pool-3.
    iii. Apply crossover process on pool-3 rules.

Fig. 4 The "Genetic Algorithm" process flow chart

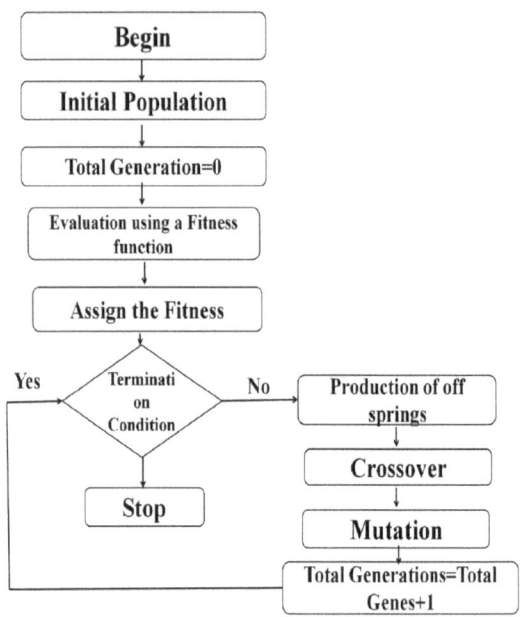

C. *Crossover operation*

The crossover operation takes two individuals as input expressions and selects a random point and interactions the sub-expression at the rear point. While implementing the crossover operation, there may be chance of having errors such as a crossover may produce individuals in which an attribute may be involved more than once and the offspring may be the same as that of the parent expression with a probability of 1. To overcome these errors, we used the below-mentioned techniques while performing crossover operation.

If two individual parents have been selected for crossover, the crossover operator is considered as follows:

1. Select one individual expression randomly among the attributes and exchange the corresponding term between them if the two individuals have same attributes
2. Select one attribute randomly from individual parent expressions, respectively, and exchange the corresponding term between the two individuals if there are not same attribute between the individual parents.

D. *Mutation operation*

Mutation operation plays a vital role in the global optimization process. This operator maintains the diversity of gene in the population and guarantees that the search is done in the absolute solution space.

Considering an individual parent "*p*" with a rule length "*n*," the mutation operator is defined as follows:

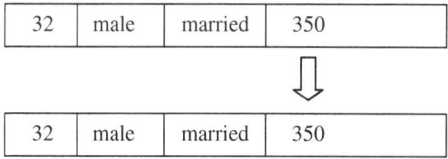

Fig. 5 Mutation with attribute modification

1. When $n = L$, select one attribute randomly from individual p and delete the corresponding term.
2. When $L > n > 1$, select one attribute arbitrarily from individual p and calculate the fitness of individual p. If $F(p) \neq 0$, then delete the consequent term or replace the attribute value by another. If $F(p) = 0$, then delete the corresponding term.
3. When $n = 1$, replace the present attribute value with another attribute on the probability.

Mutation with attribute modification:
Mutation with attribute modification is shown in Fig. 5:
Steps from mutation with attribute modification:

  i. Select a random point within the attribute range.
 ii. Form the new rule by changing the selected attribute value.
iii. Compute the fitness of the newly formed rule if the fitness is greater than the parent then add this rule to the initial population.
iv. Repeat the above process for required number of times.
 v. Mutation with attribute removal.

Figure 6 represents the chromosome before mutation and after mutation; the arrow in the Fig. 6 indicates the mutated attribute.
Sequential steps in mutation with attribute removal:

  i. Select a random point within the attribute range.
 ii. Form the new rule by removing the attribute value at the selected point.
iii. Compute the fitness of the new rule. If the fitness of the new rule is greater than the parent then add this chromosome to the initial population.
iv. Repeat this process for required number of times.

| 32 | male | married | 0 |

⇩

| 32 | male | married | 0 |

Fig. 6 The chromosome before mutation and after mutation

3.2 Stopping Criteria

The stopping criteria are as follows: If all the records in the dataset belong to the same class. If the dataset has all similar records in the attribute list. If the dataset is empty.

3.3 Performance Metrics

The most widely used performance measures in the computational intelligence techniques are accuracy, sensitivity, and specificity. The ability of the model to correctly predict the class label of previously unseen or new data is defined as accuracy. The accuracy, sensitivity, and specificity measures are shown in Table 3.

4 Results

With the performances of the GA with the traditional classification methods, we calculate the accuracy. GA approach is to construct computational intelligence techniques for the diagnosis of AIDS disease.

These rules followed to Table 2 are AIDS dataset maintaining some results are

Rule 1: Apply Rule 1 in AIDS dataset of Table 2. Calculations are TP, TN, FP, and FN.

True positive value is 4, True negative value is 18, False positive value is 0, False negative value is −2, L(no of attributes) = 4, b(length of current rule) = 2.
Sensitivity = 2.0, Specificity = 1.0, Comprehensibility = 0.666. Finally calculated Fitness Function is as follows:
Fitness Function = 2 * 1 + 0.2(0.666) = 2.1332,
Accuracy = 1.1

Table 3 Performance metrics

Measure	Formula
Accuracy	$\frac{TP+TN}{TP+TN+FP+FN}$
Sensitivity	$\frac{TP}{TP+FN}$
Specificity	$\frac{TN}{TN+FP}$

Rule 2: As followed Above Calculations Sensitivity = 3.5, Specificity = 1.1875, Comprehensibility = 0.666

Fitness Function = 4.28945,
Accuracy = 1.444.

Rule 3: As followed Above Calculations Sensitivity = 1.0, Specificity = 0.8888, Comprehensibility = 0.6666

Fitness Function = 1.022,
Accuracy = 1.28571.

Rule 4: As followed Above Calculations Sensitivity = 3.5, Specificity = 1.1875, Comprehensibility = 0.666

Fitness Function = 4.28945,
Accuracy = 1.444.

These rules are used to consider AIDS disease severity. And these rules are followed for accuracy, sensitivity, and specificity results (see Table 4).

The Clonal Selection Algorithm using to calculate accuracy, sensitivity, specificity and Fitness Function using Columns (see Fig. 7).

Table 4 Presentation of sensitivity, specificity, fitness function, accuracy of rules

Rules	Sensitivity	Specificity	Fitness function	Accuracy
Rule 1	2.0	1.0	2.1332	1.1
Rule 2	3.5	1.1875	4.28945	1.44
Rule 3	1.0	0.8888	1.022	1.28571
Rule 4	3.5	1.187	4.28877	1.4

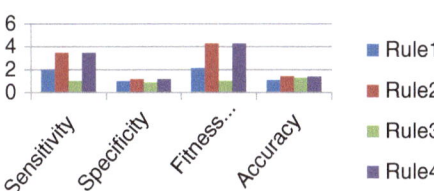

Fig. 7 Presentation of columns sensitivity, specitivity, fitness function, accuracy of rules

5 Conclusion

The proposed GA of Fitness Function using Good methodology using developed is found to be effective in diagnosis of AIDS Disease clinical condition. Patient Data is considered as experimental validation of the proposed system. The rule based on Determines the AIDS condition based on the High or Low values of CD4 cell count. GA using developed techniques has been used for identification of AIDS disease severity.

References

1. de Castro LN, Timmis J (2002) Artificial immune systems: a new computational intelligence approach. Springer, Berlin
2. Dasgupta D, Forrest S (1995) Novelty detection in time series data using ideas from immunology. In: Presented at the Proceedings of the 5th international conference on intelligent systems, Reno, Nevada
3. Greensmith J, Aickelin U, Cayzer S (2005) Introducing dendritic cells as a novel immune-inspired algorithm for anomaly detection. In: Jacob C, Pilat M, Bentley P, Timmis J (eds) Artificial immune systems. Springer, Berlin, pp 153–167
4. Igawa K, Ohashi H (2009) A negative selection algorithm for classification and reduction of the noise effect. Appl Soft Comput 9:431–438
5. Holland JH (1975) Adaptation in natural and artificial systems. University of Michigan Press, Ann Arbor, p 183 re-issued by MIT Press 1992
6. Bhattacharjya RK (2012) Introduction to genetic algorithms. Guwahati, IIT, p 12
7. Han J, Kamber M, Pei J (2011) Data mining: concepts and techniques. Third. s.l.: Morgan Kaufmann, Los Altos
8. Varma KV, Rao AA, Sita Mahalaxmi T, Nagaswara Rao PV, Narasimha Rao Kandula A computational intelligences technique for effective diagnosis of diabetes disease using genetic algorithm
9. Al-Maqaleh BM, Shahbazkia H (2012) A genetic algorithm for discovering classification rules in data mining. Int J Comput Appl 41(18):40–44 (0975–8887)
10. Pradhan MA, Bamnote GR, Tribhuvan V, Jadhav K, Chabukswar V, Dhobale V A genetic programming approach for detection of diabetes. Int J Comput Eng Res (ijceronline.com) 2(6)
11. Permann MR (2007) Genetic algorithms for agent-based infrastructure interdependency Modeling and analysis INL/CON-07-12317, SpringSim
12. Carvalho DR, Freitas AA An immunological algorithm for discovering small-disjunct rules in data mining
13. Dharmaiah Devarapalli, Allamapparao, Amit Kumar,G R Sridhar (2013) A novel analysis of diabetes mellitus by using expert system based on brain derived neurotrophic factor (BDNF) levels. Helix 1:251–256. ISSN–2319, 5592

Author Biographies

Dr. Devarapalli Dharmaiah is currently working as an Associate Professor in Computer Science and Engineering department, Vignan's Institute of Information Technology, Visakhapatnam and has the teaching and research experience of about 10 years. He also guided various dissertation works for both UG and PG students of VIGNAN'S IIT and other Colleges. He has taught various subjects of Computer Science and Applications for both of UG & PG students such as C programming, Data Structures, Java, Operating Systems, Compiler Design, Linux, UNIX and Bioinformatics. He has a very good reputation among the students and faculty community for his proficiency in subjects. He is a lifetime member of CSI. He has published many papers in National, International conferences and leading Journals. His areas of interest are Bioinformatics, Neural Networks, Data Mining and Computer Networks.

Mr. Panigrahi Srikanth is currently pursuing his 2 years of M.Tech (SE) in Department of Computer Science and Engineering at Vignan's Institute of Information Technology, Visakhapatnam. He received his B.Tech (IT) from Gokul Institute of Technology and Sciences, Piridi, Bobbili, Vizia-nagaram, A.P. His current research is on Medical Diagnosis (Heart, Hepatitis, Diabetes, Thyroid and Cancer Diseases). His areas of interest include Bioinformatics, Data Mining and Information Security.

A Novel Clustering Approach Using Hadoop Distributed Environment

Nagesh Vadaparthi, P. Srinivas Rao, Y. Srinivas and M. Athmaja

Abstract Nowadays, information retrieval plays a vital role by allowing users to retrieve documents of their interest based on relevance score. Such systems can be implemented either in distributed systems or parallel systems to achieve high throughput. If such kind of framework is deployed in a cloud, grouping of relevant documents is essential to retrieve documents of interest. Hence, an efficient and scalable clustering is required to process huge volume of documents. To handle huge documents and to provide scalability while processing Apache Hadoop is efficient with its powerful feature map reduce. Hence, in this paper, a novel approach is proposed that is capable of clustering bulk data with high throughput. This paper also demonstrates the need of parallel caching approach for obtaining effective results.

Keywords Data clustering · Parallel computing · Hadoop · HDFS · MapReduce

1 Introduction

In this fast pace world of ever changing technology, we have been drowning in information. We are generating and storing massive quantities of data with the proliferation of devices on our networks, and we have seen an amazing growth in a diversity of information formats and data.

N. Vadaparthi (✉) · P. Srinivas Rao
MVGR College of Engineering, Vizianagaram, India
e-mail: itsnageshv@gmail.com

P. Srinivas Rao
e-mail: psr.sri@gmail.com

Y. Srinivas
GIT, GITAM University, Visakhapatnam, India
e-mail: sriteja.y@gmail.com

M. Athmaja
Tata Consultancy Services, Hyderabad, India
e-mail: athmaja.mvgrit@gmail.com

© The Author(s) 2015   113
N.B. Muppalaneni and V.K. Gunjan (eds.), *Computational Intelligence Techniques for Comparative Genomics*, Forensic and Medical Bioinformatics,
DOI 10.1007/978-981-287-338-5_9

Over the course of the past decade, many technologies have promised to help with the processing and analyzing of the vast amounts of information [1] we have, and most of these technologies have come up short. We know this because as programmers focused on data, we have tried it all. Many approaches have been proprietary, resulting in vender lock-in. Some approaches are promising but couldnot scale to handle large datasets and many were hyped up so much that they couldnot meet expectations, or they simply were not ready for prime time.

When Apache Hadoop [2] entered the scene, however, everything was different. Hadoop is an open source that had already found incredible success in massively scalable commercial applications. Based on a MapReduce [3, 4] algorithm that enable us to bring processing to the data distributed on a scalable duster of machines, we have found much success in performing complex data analysis in ways that we havenot been able to do in past.

There was various numbers of methods of data analysis in the field of data mining, pattern recognition, image processing, etc. Out of the existing methods, K-Means is widely used. But clustering becomes more and more complex when the process is done for large-scale datasets. The time complexity of K-Means algorithm is $O(NKD)$ where N is the number of objects, D number of iterations and k number of clusters.

But the disadvantage with K-Means algorithm is k should be initialized and the result varies with the value of k. Another disadvantage is it requires additional space to store the data, and also for a given initial seed set of cluster centers, it generates the same partition of the data irrespective of the order in which the patterns are presented. Also, it doesnot necessarily find the most optimal [5]. It is sensitive to the order of data input [6].

Hence, there is a need for an enhance algorithm that can minimize the above disadvantage. Therefore, this paper introduces a novel and efficient technique when the dataset is large. In this paper, we propose a new technique that includes FCM with canopy algorithm. However, the implementation of FCM with canopy on distributed computing yields better results.

The rest of the sections are organized as follows: Sect. 2 describes the architecture of the proposed method, Sect. 3 demonstrates about clustering process using canopies, Sect. 4 elaborates Fuzzy C-Means algorithm for clustering the semi-clustered groups, and the results of the proposed method have been discussed in Sect. 5. Finally, Sect. 6 concludes the paper.

2 Architecture of the Proposed Method

In the proposed architecture, the data available are initially fed as input to find the approximate process of clustering to canopy technique and in the next step, the points are assigned to canopy. After obtaining the initial clusters, the obtained cluster groups are fed to FCM algorithm and the final clusters are obtained. The Fig. 1 below demonstrates the different steps in the proposed architecture.

Fig. 1 Architecture of the proposed method

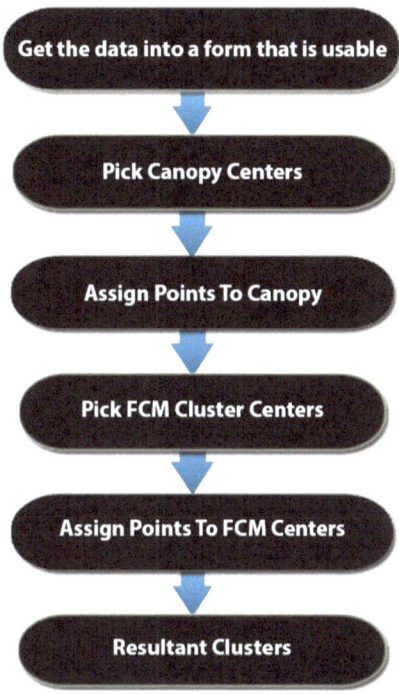

3 Clustering Using Canopies

In this algorithm, the computations are reduced by initially partitioning approximately the data into overlapping subsets, and then, it measures the distances among pairs of data points that belong to a common subset [7].

In this technique, two different sources of information are being used for clustering the items (1) cheap and approximate similarity measure and (2) an expensive and accurate similarity measure [7]. All points are represented as a point in multidimensional feature space. Canopy is often used as an initial step to significantly reduce the more expensive distance measure by ignoring points that are outside the initial canopies. But this algorithm is being implemented under parallel environment. The strategy for parallelization of the canopy clustering is as follows:

1. The input data are massaged to required format.
2. Canopy clustering is performed on the input set by each mapper function and gives output as canopy centers.
3. Now, the final canopy centers are obtained from the clusters produced by the reducer function.
4. Finally, the obtained points are clustered to final canopies.

Figure 2 demonstrates the procedure of canopy clustering technique. With the utilization of the above parallel approach, the canopies are created. But, as the

Fig. 2 Procedure for canopy clustering algorithm

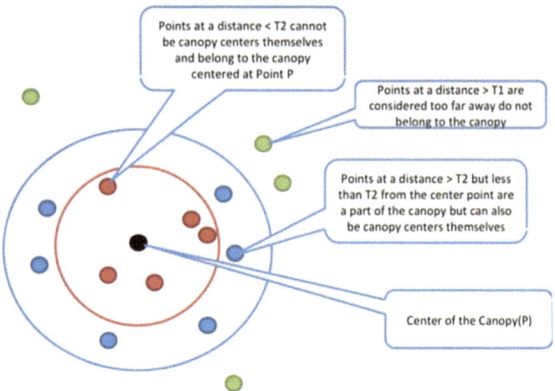

distance measure used to create canopies are approximate, the obtained clusters are not accurate. Hence, for obtaining much accurate clusters, the obtained output canopies are fed as input to the fuzzy C-means algorithm. Earlier literature addresses the other algorithms [8] like K-Means efficiently performs clustering by finding good initial starting point, but is not efficient when the number of clusters is large. Another disadvantage of K-Means is that it works effectively only for smaller datasets.

Hence, in this paper, we have utilized the Soft-Fuzz (Fuzz C-Means) algorithm for efficient clustering. The algorithm for Fuzzy C-Means clustering is presented in the next section.

4 Fuzzy C-Means Clustering Algorithm

The Fuzzy C-Means clustering applies fuzzy partitioning process so that the data point might belong to all groups having different membership grades ranging from 0 to 1 and is iteratively performed. Finding cluster centers that minimize dissimilarity function is the main aim of Fuzzy C-Means algorithm. For fuzzy partitioning, the membership matrix (U) is being randomly initialized as per Eq. (1).

$$\sum_{i=1}^{c} u_{ij} = 1, \quad \forall j = 1, \ldots, n \tag{1}$$

The dissimilarity function used is given as

$$J(U, c_1, c_2, \ldots, c_c) = \sum_{i=1}^{c} J_i = \sum_{i=1}^{c} \sum_{j=1}^{n} u_{ij}^{m} d_{ij}^{2} \tag{2}$$

where $0 < u_{ij} < 1$

c_i centroid of cluster i;

d_{ij} Euclidian distance between ith centroid (ci) and jth data point

$m \in [1, \infty]$ is a weighting exponent.

To attain minimum dissimilarity function, there exist two conditions which are given in Eqs. (3) and (4).

$$c_i = \frac{\sum_{j=1}^{n} u_{ij}^m x_j}{\sum_{j=1}^{n} u_{ij}^m} \qquad (3)$$

$$u_{ij} = \frac{1}{\sum_{k=1}^{c} \left(\frac{d_{ij}}{d_{kj}}\right)^{2/(m-1)}} \qquad (4)$$

This algorithm ascertains the following steps.

Step-1: Randomly initializing the membership matrix (U) which has constraints in Eq. (1).

Step-2: Calculating centroids (ci) by using Eq. (3).

Step-3: Computing dissimilarity between the data points and the centroids using Eq. (2).

   If the threshold value is below the previous iteration, then stop the process, else continue.

Step-4: Compute the new "U" using Eq. (4) and go to Step 2.

On iteratively updating the membership grades and the cluster centers for each data point, Fuzzy C-Means iteratively moves cluster centers to the "right" location within the dataset. Fuzzy C-Means shall not ensure that it converges to optimal solution for the reason being the cluster centers (centroids) are initialized using "U" which are randomly initialized (Eq. 3).

The Performance of algorithm depends on initial centroids. There are two ways for robust approach viz.,

1. Determining all the centroids by using algorithm.
2. Execute FCM iteratively starting with initial centroids.

However, as the canopy clustering technique is utilized as a pre-processing step for Fuzzy C-Means algorithm, the results obtained by the application of FCM has converged to an optimal solution.

5 Results and Discussions

The experimentation has been performed on FOUR (04) nodes with 1-master node and 3-slave nodes. The configuration of the master node is Intel Core2 Duo with 2 GB RAM and the slave nodes with configuration as Intel Core2 Duo with 1 GB RAM.

Table 1 Comparison of classifiers

Classifier	# Correct classified samples	Correct classification rate (%)
K-Means with canopy	172	79.2
Fuzzy C-Means with canopy	186	84.8

We have made an attempt to show the difference in execution times on single node and multi nodes along with the comparison between Hard Clustering (K-Means) and Soft Clustering (Fuzzy C-Means) techniques. The experimentation is carried out using different size data points of 1,000, 100,000, 1,000,000, and 10,000,000 records. The comparison is also made to demonstrate the number of classified samples and the correctness of the classification with respect to both K-Means and Fuzzy C-Means algorithms with canopy in Table 1. With the results in Table 1, it is clear that the FCM with canopy is effective than K-Means with canopy. The experimentation is done on Ubuntu 12.10, Hadoop 0.20.1 and Mahout 6 environment using Java7.

In Fig. 2, the experimentation is done using different size datasets and the time to complete the clustering process is analyzed. The graph shows the comparison between K-Means with canopy and Fuzzy C-Means with canopy techniques. It is observed that the time taken to cluster the data is almost equal for smaller datasets. But as the size of the dataset increases, decrease in the time taken is reduced for Fuzzy C-Means technique than the K-Means techniques.

Hence, from the Fig. 3 and Table 1, it is evident that the proposed method Fuzzy C-Means with canopy technique is more efficient than the K-Means with canopy technique.

Fig. 3 Graph showing the number of documents versus time required to process

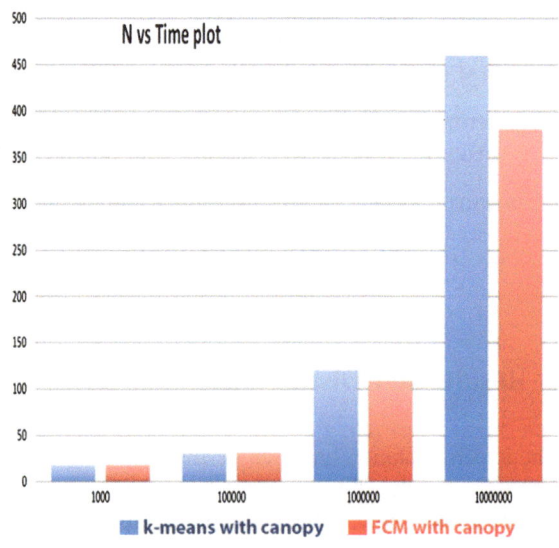

6 Conclusion

In this paper, we made an attempt to implement Fuzzy C-Means with canopy and compare the results with K-Means with canopy clustering technique. The experimentation was done with 4 nodes of which 1-master node and 3-slave nodes. It is evident from the obtained results that the time consumed for smaller datasets is almost same for both the algorithms and a drastic decrease is observed in the time consumed to cluster data by utilizing FCM with canopy in Hadoop environment. We have observed that the parallel Hadoop implementation is effective for large datasets when a huge computationally intensive application is required.

References

1. Lynch C (2008) Big data: how do your data grow? Nature 455(7209):28–29
2. Ye K et al (2012) vHadoop: a scalable hadoop virtual cluster platform for mapreduce-based parallel machine learning with performance consideration. In: IEEE international conference on cluster computing workshops, pp 152–160
3. Dean J et al (2008) MapReduce: simplified data processing on large clusters. Commun ACM 51 (1):107–113
4. White T (2010) Hadoop: the definitive guide. Yahoo Press
5. Vadaparthi Nagesh et al (2011) Segmentation of brain MR images based on finite skew gaussian mixture model with fuzzy C-Means clustering and -EM algorithm. Int J Comput Appl 28(10):18–26
6. Sabena S et al (2011) Image retrieval using canopy and improved K mean clustering. In: International conference on emerging technology trends (ICETT) 2011, pp 15–19
7. McCallum A et al (2011) Efficient clustering of high-dimensional data sets with application to reference matching. White papers
8. Bradley PS et al (1998) Scaling clustering algorithms to large databases. In: Proceeding of 4th international conference on knowledge discovery and data mining (KDD-98). AAAI Press, Menlo Park

Framework for Evaluation
of Programming Language Examinations

Himani Mittal and Syamala Devi Mandalika

Abstract Recent advancements in the field of e-learning and virtual learning have changed the face of education. An important part of learning process is evaluation of student learning through examinations. This paper suggests a framework for evaluation of computer science practical examinations. The framework is implemented using Java programming language and many open source tools and libraries. The developed framework performs evaluation in four steps. The first step is compiler output interpretation in which the false errors generated by compiler are eliminated and only actual errors are reported. In the second step, unit testing of compiled programs is done. In third step, software metrics like lines of code, lines of comment, McCabe's cyclomatic complexity, and number of modules are calculated for the programs. Finally, the semantic similarity of student programs is checked against the model program. The implemented framework is tested on student programs, and the accuracy of results is satisfactory. This framework will be helpful in efficiently evaluating student programs in practical examinations. It works for C, C++, and Java programming languages.

Keywords Practical evaluation · C · Java · C++ · Program evaluation

1 Introduction

Teaching and learning process is undergoing a transition. Virtual learning and e-learning platforms have changed the face of classroom teaching. Examinations are an important part of learning process which requires to be computerized. In traditional examination system, the student writes his program in a programming

H. Mittal (✉) · S.D. Mandalika
Department of Computer Science and Applications, Panjab University, Chandigarh, India
e-mail: research.himani@gmail.com

S.D. Mandalika
e-mail: syamala@pu.ac.in

N.B. Muppalaneni and V.K. Gunjan (eds.), *Computational Intelligence Techniques for Comparative Genomics*, Forensic and Medical Bioinformatics, DOI 10.1007/978-981-287-338-5_10

language and tests. At the end, examiner personally checks each student program and evaluates. This is time consuming and repetitive task. Often the evaluation is not perfect. Also there is scope for bias and complaints from students. Hence, there is a need to automate the evaluation process. The evaluation process includes checking of the compilation errors, testing of logic and results and efficiency of the program. In this paper, a framework is developed for performing practical examination/evaluation covering the above-mentioned evaluation process. The use of computerized tools can reduce the limitations of manual process. Computerized evaluation ensures uniformity of marking scheme because it has the same inference mechanism for checking all the answers/programs. Automation can ensure speedy result declaration.

Practical examinations for computer science education include project work, assignments, and programming experiments. The already available tools focus only on the evaluation of programming assignments and competitions. The framework developed in this paper can be used for classroom environment, for assessment, for competitions, and also for conducting final examination of students at the end of semester.

The paper is organized as follows. The review of related work in practical evaluation tools is given in Sect. 2. The framework and its working are given in Sect. 3. The list of algorithms, tools, and techniques used for evaluation are discussed in Sect. 4. Conclusions and scope for future work is mentioned in Sect. 5.

2 Review of Related Work

Jackson and Usher [1] developed a computer-based evaluation tool called ASSYST. This tool evaluates computer science practical assignments of C and ADA language. It only assists in evaluation. It assesses computer programs on four parameters: correctness, style, efficiency, and complexity. RoboProf [2] is an online teaching system based on the World Wide Web technology. RoboProf teaches syntax and structure of C++ programming language to students and assesses the exercises students solve. BOSS [3] is an online submission system, which provides features for course management and automated testing. It could evaluate C, C++, and Java programs. It makes use of JUnit for testing. Mooshak tool was developed in 2003 by Leal and Silva [4]. Mooshak is online programming contest managing software. It helps in conduct, answer evaluation, result declaration, and feedback. The languages supported by default are C, Pascal, C++, and Java. In 2009, García-Mateos and Fernández-Alemán [5] and Montoya-Dato et al. [6] used Mooshak for algorithms and data structures course.

Douce et al., in 2005 [7], developed Automated System for Assessment of Programming (ASAP) project which fits into the e-learning framework as an evaluation tool. Mandal et al. discussed, in their paper published in 2006 [8], the architecture of an automatic program evaluation system. This system can handle only C programs. Their approach is to perform white box test, instead of, black box or gray box testing. SQLify (2007) [9] is to assist students learn SQL. It facilitated writing test queries against databases; receive immediate feedback which is more informative than what can be offered by DBMS. In 2010, Zhang and Ke [10] also proposed a design for SQL Paperless Examination System.

In 2008, Farrow and King [11] used an online programming examination to assess undergraduates who had learned Java for two terms. It uses BLUEJ and JUnit software. In 2010, Skupas [12] wrote a paper on feedback improvement in automatic program evaluation systems. He says that some programming errors should not influence the final score heavily. Black box testing alone cannot ensure this. A typical approach is to use semi-automatic systems for programming test evaluation. It proposes two grading steps. The first step is responsible for student program output format checking. The second step relates to output correctness. In 2011, Fernández-Alemán [13] proposed an automated assessment tool for programming courses. It extends the functionality of Mooshak by adding an external static marker.

In paper by Zhenming et al. [14] and later extended by Zhang et al. [15], tool was developed to measure operating skills of the students and their proficiency in using software like MS Word, MS Excel, MS Power Point, and Internet familiarity. An overview of features required by practical evaluation tools is given in [16]. For Evaluation of GUI-based student programs English [17] proposed a tool called JEWL for Java GUI programs.

The available practical evaluation tools are used for different purposes. Some tools are used for assignment evaluation and classroom teaching. Some are used for conducting online programming contests and evaluating contestants' programs. Programming languages evaluated by these tools are C, C++, FORTRAN, SQL, Pascal, and Java (all tools do not evaluate all languages). The technique used in all the tools for evaluation is black box testing. A more holistic approach is required for evaluating the student programs. Apart from black box testing, the program quality and semantic similarity of program with the algorithm needs to be established.

3 Practical Evaluation Framework

The students submit their programs to the server where they are stored in a database. Then, each program is read from the database and it passes through the following four stages.

(a) Compilation stage: In the first stage, the syntax of the program needs to be checked. The students who pass in this stage will go to next stage. The syntax is already checked by the programming language compilers. However, there is a problem in the output. If the programmer makes a small mistake like missing semicolon, missing bracket, or something similar, then the compiler shows errors in all the lines following the erroneous line. In order to remove prop-agated errors, in this framework, the compiler output is given as input to a small program and it points out all the redundant errors reported. Some marks are given to the student if the actual errors reported are less.

(b) Unit testing stage: After the program is successfully compiled, it needs to be tested for functionality by performing unit testing of the programs. In unit testing, it is checked that whether all the modules are working as desired or not. It is performed by white box testing of the student program. The test cases are provided by the examiner. The test cases are run on model answer to find the expected output. Then, the student program is evaluated for the same test cases. If there is a match between the expected output and student program output, it is a success. For ensuring the success of this functionality, the signatures (number of arguments and return value) of student functional units must match the model program.

(c) Metrics calculation stage: After the program is tested for correctness, its complexity and style need to be evaluated. The lines of Code (KLOC), lines of Comment, McCabe's cyclomatic complexity, and number of Modules are calculated. The metric values of student program are matched with that of model program. If the number of lines of code is more in student program, then it is negative as lesser lines are solving the problem. If lines of comment are less, then it is negative as program is not self-explanatory. If number of modules is less, then coupling and cohesion are not ensured. If McCabe's cyclomatic complexity is more, then program is unnecessarily made complex.

(d) Semantic similarity stage: The semantic similarity of the student program to the standard program needs to be established. This is done by creating system dependency graphs (SDG) of student program and standard program. Both the graphs are compared for similarity. Exact matches in the graph are not required. The number of modules, their internal nodes, and dataflow are measured.

3.1 Integration of Evaluation Stages into Evaluation Framework

The diagram showing the different stages of framework is given in Fig. 1. The stages of evaluation mentioned above are integrated using Java program into one framework. The integration Java program invokes several tools and libraries, captures their output and interprets it.

The total marks to be awarded for the practical question can be given based on the four stages. The suggested evaluation strategy can be as follows. The first part is given to the student if the program compiles successfully and passes the functional

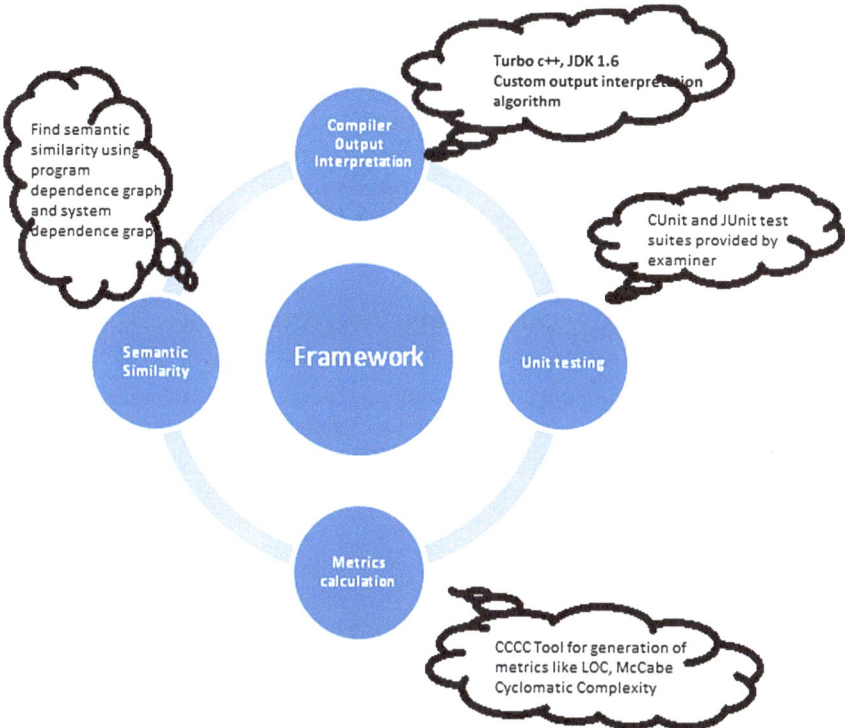

Fig. 1 Steps in framework

testing stage. If the program does not pass the functional testing phase, then marks are given based on formula:

$$(\text{actualErrors}/\text{reportedErrors}) * (\text{one fourth of total marks}).$$

The second part is given to the student based on test case passing. The third part of marks is given on the basis of efficiency and complexity metrics. The last part is given on the basis semantic similarity between student program and model program.

4 Implementation of the Framework

The framework is developed on Microsoft Windows 8 environment using Java programming language and Java Agent Development Environment (JADE) [18]. The developed framework can be used for programming languages C, C++, and Java.

Tools and open source libraries used for evaluation are:

- Turbo C++ compiler, DosBox, and Java development kit (JDK 1.6) for compiler output generation.
- Interpretation of compiler output is done using custom algorithm.
- CUnit [19] and JUnit [20] libraries are used for unit testing of student program.
- CCCC Tool [21] is used for software metric calculation.
- The FramaC Tool [22] and SDG Java library [23] are used for SDG generation.

4.1 Algorithm for Compiler Output Interpretation

The compiler output is interpreted for removing redundant error reports. Sometimes due to single error, many dummy errors are reported. The method followed to remove the redundant error lines is as follows. The line number and error string are compared with the line number and error sting of subsequent error, if they are same, then the error is ignored.

4.2 Testing the Complexity of Program

The CCCC Tool generates a HTML file containing several metric values for the student program. The numeric values are extracted from this file using Java packages: org.w3c.dom.Document, org.w3c.dom.*, javax.xml.parsers. Document-BuilderFactory, and javax.xml.parsers.DocumentBuilder. The framework extracts the values from this HTML file for student program as a whole and for each functional unit.

4.3 Logic Testing Using Dependency Graphs

The FramaC Tool and SDG Java library are used to generate SDG and program dependency graph (PDG) from student program code and model program code. There are traversal functions defined for depth-first Traversal and breadth-first traversal of these graphs on the basis of controlflow and dataflow. Both of these traversals are used to assess the similarity of any two programs.

4.4 Testing and Results

For testing purposes, a class test was conducted manually and the student answers were collected. It has been found that framework evaluates the student programs on all the four parameters and results are satisfactory.

5 Conclusions and Scope for Future Work

The developed framework will be helpful for the examiner in the evaluation of student practical examinations with minimum human intervention. The use of computerized tools can reduce the limitations of manual process. Computerized evaluation ensures uniformity of evaluation, speedy, and transparent results.

The size of programs evaluated by the software is from small practice questions to single file programs. The scope of the current system is limited to C, C++, and Java programming languages.

The framework can be improved to check the student programs for viruses by integrating with some anti-virus program. It can be improved to handle multi-file projects. The framework can be extended to multi-agent technology including student agent and examiner agent. The examiner agent can also be used for conducting the written objective test of the students that may be a part of the practical examination.

References

1. Jackson D, Usher M (1997) Grading Student Programs using ASSYST. In: Proceedings of the 28th SIGCSE technical symposium pp 335–339. doi:10.1145/268084.268210
2. Daly C (1999) RoboProf and an introductory computer programming course. In: Proceedings of the 4th annual SIGCSE/SIGCUE on innovation and technology in computer science education, Krakow, pp 155–158, 27–30 June 1999. doi:10.1145/384267.305904
3. http://www.boss.org.uk/
4. Leal JP, Silva F (2003) Mooshak: a web-based multi-site programming contest system. J Software—Pract Experience 33(6):567–581. doi:10.1002/spe.522
5. García-Mateos G, Fernández-Alemán JL (2009) A course on algorithms and data structures using on-line judging. In: Proceedings of the 14th annual ACM SIGCSE conference on innovation and technology in computer science education 41(3):45–49. doi:10.1145/1505496. 1562897
6. Montoya-Dato FJ, Fernández-Alemán JL, García-Mateos G (2009) An experience on Ada programming using on-line judging. In: Proceedings 14th international conference on reliable software technologies, pp 75–89. doi:10.1007/ 978-3-642-01924-16
7. Douce et al (2005) A technical perspective on ASAP—automated system for assessment of programming. In: Proceedings of the 9th computer assisted assessment conference
8. Mandal AK, Mandal C, Reade CMP (2006) Architecture of an Automatic program evaluation system. In: Proceedings of CSIE
9. Raadt M, Dekeyser S, Lee TY (2007) A student employing peer review and enhanced computer assisted assessment of querying skills. Inform Educ 6(1):163–178
10. Zhang G, Ke H (2010) SQL paperless examination system design. In: 2010 Second international conference on computer modeling and simulation, IEEE, 3:475–478. doi:10. 1109/ICCMS.2010.468
11. Farrow M, King PJB (2008) Experiences with online programming examinations. IEEE Trans Educ 51(2):251
12. Skupas B (2010) Feedback improvement in automatic program evaluation systems. Inform Educ 9(2):229–237
13. Fernández Alemán JL (2011) Automated assessment in a programming tools course. IEEE Trans Educ 54(4):576–581

14. Zhenming Y, Liang Z, Guohua Z (2003) A novel web-based online examination system for computer science education. In: Proceedings of international conference on frontiers in education, IEEE, New York
15. Zhang L, Zhuang YT, Yuan ZM, Zhan GH (2006) A web-based examination and evaluation system for computer education. In: proceedings of the sixth international conference on advanced learning technologies, IEEE, New York
16. Hollingsworth J (1960) Automatic graders for programming classes. Commun ACM 3 (10):528–529. doi:10.1145/367415.367422
17. English J (2004) Automated assessment of GUI programs using JEWL. In: Proceedings of the 9th annual SIGCSE conference on innovation and technology in computer science education. pp 131–141
18. http://jade.tilab.com/
19. http://cunit.sourceforge.net/
20. http://junit.org/
21. http://sourceforge.net/projects/cccc/
22. http://frama-c.com/
23. SDG library—http://www4.comp.polyu.edu.hk/~cscllo/teaching/SDGAPI/

An Efficient Data Integration Framework in Cloud Using MapReduce

P. Srinivasa Rao, M.H.M. Krishna Prasad and K. Thammi Reddy

Abstract In Bigdata applications, providing security to massive data is an important challenge because working with such data requires large scale resources that must be provided by cloud service provider. Here, this paper demonstrates a cloud implementation and technologies using big data and discusses how to protect such data using hashing and how users can be authenticated. In particular, technologies using big data such as the Hadoop project of Apache are discussed, which provides parallelized and distributed data analyzing and processing of petabyte of data, along with a summarized view of monitoring and usage of Hadoop cluster. In this paper, an algorithm called FNV hashing is introduced to provide integrity of the data that has been outsourced to cloud by the user. The data within Hadoop cluster can be accessed and verified using hashing. This approach brings out to enable many new security challenges over the cloud environment using Hadoop distributed file system. The performance of the cluster can be monitored by using ganglia monitoring tool. This paper designs an evaluation cloud model which will provide quantity related results for regularly checking accuracy and cost. From the results of the experiment found out that this model is more accurate, cheaper and can respond in real time.

Keywords Big data · Hadoop · MapReduce · Cloud computing · Accuracy · Consumption

P. Srinivasa Rao (✉)
Department of Computer Science Engineering, MVGR College of Engineering, Vizianagaram, India
e-mail: psr.sri@gmail.com

M.H.M. Krishna Prasad
Department of Computer Science Engineering, Jawaharlal Nehru Technological University, Kakinada, India
e-mail: krishnaprasad.mhm@gmail.com

K. Thammi Reddy
Department of Computer Science Engineering, GITAM University, Visakhapatnam, India
e-mail: thammireddy@yahoo.com

© The Author(s) 2015 129
N.B. Muppalaneni and V.K. Gunjan (eds.), *Computational Intelligence Techniques for Comparative Genomics*, Forensic and Medical Bioinformatics, DOI 10.1007/978-981-287-338-5_11

1 Introduction

With increasing amount of digital data, the term big data has been popular nowadays; this describes the experimental growth and usage of structured data as well as semi-structured data. To work with this much of data, there is in need to find a new methodology in a way to produce accurate results [1]. Nowadays, if we look at the big data market forecast, it will be grown by 2017 nearly US$50–60 Billions. The jobs filled in big data are more when compared to other areas. Processing of such big data in cloud requires an efficient, scalable, and effective processing and computational tool such as Hadoop. This open source framework together with MapReduce has created an evolution in large-scale computing. The four highlighted features of this are the following: scalability, low cost, ease of usage, and fault tolerance. Effective solutions were offered by Hadoop to manage and to process this massive data [2]. The customization of MapReduce to analyze massive data makes Hadoop the most preferred and admired technology for handling big data.

Cloud computing has become a hot topic both in research and in industry. It is any environment which is created in a user's system from an online application on the cloud and works through a Web browser. Cloud service refers to providing computing resources through Net remotely. The main cloud computing services are IAAS, PAAS, and SAAS. Cloud computing allows individual and organizations to use resources that are managed by service providers at remote locations. The main characteristics of cloud service are shared resources that are provided dynamically with pay-as-you-use. When making decisions on deploying or outsourcing this application into cloud computing-related solutions, security has always been a most important concern [3, 4]. Cloud computing is becoming popular because of its high reliability, availability, and importantly its low cost; with these flexibilities, many cloud storage services have been deployed. But drawback is performance and security cannot be guaranteed when data are stored over public cloud.

Data distributors will handle the data to trusted third parties, and hence, there is a chance of leakage. Some organizations apply information security only in terms of their network from intruders but with increasing amount of sensitive data as the growth of the organization leads to increase in number of data points, and sometimes, accidental or even premeditated data leakage from within the organization has become painful [5, 6].

1.1 Contribution

This paper aims to develop non-cryptographic cloud model for cloud computing-based applications for addressing the authentication and integrity. In this paper, the non-cryptographic algorithm used was the improved version of Fowler–Noll–Vo hash function FNV-1a [7]. Since it was a non-cryptographic algorithm, it uses only multiplication and XOR operations repeatedly on the octets of data [8]. And hence, the algorithm is fast and uses very less resources.

1.2 Organization of the Paper

This paper addresses integrity and authentication that must be provided to the data that are communicated between Hadoop machines, which is distributed infrastructure. And the remaining paper is arranged as follows: The related work is discussed in Sect. 2, the proposed model and its architecture is given in Sect. 3, the Methodology used is given in Sect. 4, the results of experiment are given in Sect. 5, and at last, in Sect. 6, Conclusions and Future work is given.

2 Related Work

Karthik Kambatla et al. [1] introduced a Hadoop that uses the cloud to access with a model called pay-as-you-use. Zhao et al. [9] and Wang et al. [10] introduced a method called G-Hadoop mechanism, and Rupesh and Chitre [6] presented a report on Data Leakage and Detection of Guilty Agent. Caballer et al. [11] have presented code cloud architecture and a dependent platform that supports to execute set of scientific applications which is dealt with different programming models on cloud environment. AL-Saiyd and Sail [12] designed a cloud computing security development life cycle model. Dillibabu et al. [13], proposed model for sending the data by splitting it using Merkle Hash Tree Algorithm and uses verification process. Glenn and Phong developed the hashing algorithm called FNV and Landon Curt Noll, improved the algorithm [7, 8]. Mounika et al. [14] proposed a model called a fully distributed load rebalancing algorithm, which is used to overcome load imbalance problem. To provide security, this paper proposes a new methodology where it will be done in two-stage Hadoop, which includes data integrity also.

3 Methodology

A simple Hadoop cluster framework is used to develop efficient cloud monitoring framework (ECMF). It is implemented with an efficient data integrity technique FNV to provide authorization and integrity of the data that have been published by the user in cloud environment. Following Fig. 1 shows the model architecture of the system used in this paper.

In the above figure, at any data node, user may transfer data to any data node present in cloud. In the above figure, the idea of transferring data securely from any node to any node was displayed. The process of generating message digest for the data that has been stored in cloud will be discussed in Sect. 3.2.

Fig. 1 Hadoop cluster
architecture with FNV
algorithm

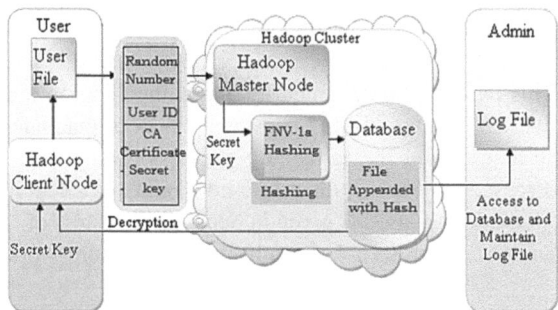

3.1 Client Authentication

The modified ECMF also consists of a single name node (NN), multiple data node (DNs), and a secondary name node (SNN) for taking backup of name node at each checkpoint. Following Fig. 2 shows client request flow.

3.2 FNV-1a Hashing Algorithm

FNV-1a is an upgraded algorithm of FNV-1. FNV is hash algorithm created by Glenn, Landon, and Phong, and it is a not cryptographic and hence works fast. **FNV** hashes are designed to be fast at the same time rate of the collisions when compared to other algorithms is very low. These hashes are widely used in many applications like DNS, applications used in cloud model, data lookup functions, database indexing hashes, log file splitting, etc.

Present available versions of FNV-1a crate a nonzero offset, and come in 32-, 64-, 128-, 256-, 512-, and 1,024-bit.

The main working of the FNV-1a algorithm is as follows:

$H = Offset$
for every data_octate
$H = H \ xor \ data\_octate$
$H = H \ *prime$
return H

Fig. 2 Client's request flow
in the cloud

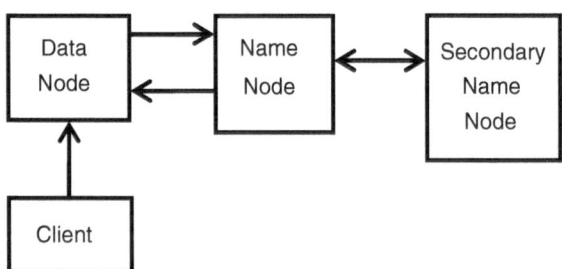

Here, *H* stands for *hash*, offset is initial vector that has to take at first, *data_octate* is input data split of an octate size, and *prime* stands for selected FNV prime.

There is FNV-1 also available; only difference between it and FNV-1a hash is that the order of the multiply and xor will be altered, and p*rime* and *offset* used by both of them are same.

3.3 Internal Operation

At a data node, if the user is authorized to access the cloud, he can make a data transfer, where the data can be spitted into parts and for each, a message digest will be generated which is called child digest; all child digests will be added to master digest file and are encrypted together with the original data of the owner and sent back to him when he requests, which is shown in Fig. 3.

Since only the latest checkpoint and operation log related are only needed, the older ones can be deleted. The operation log and checkpoints are replicated to provide for reliability by Secondary name node. The ECMF significantly improves its robustness.

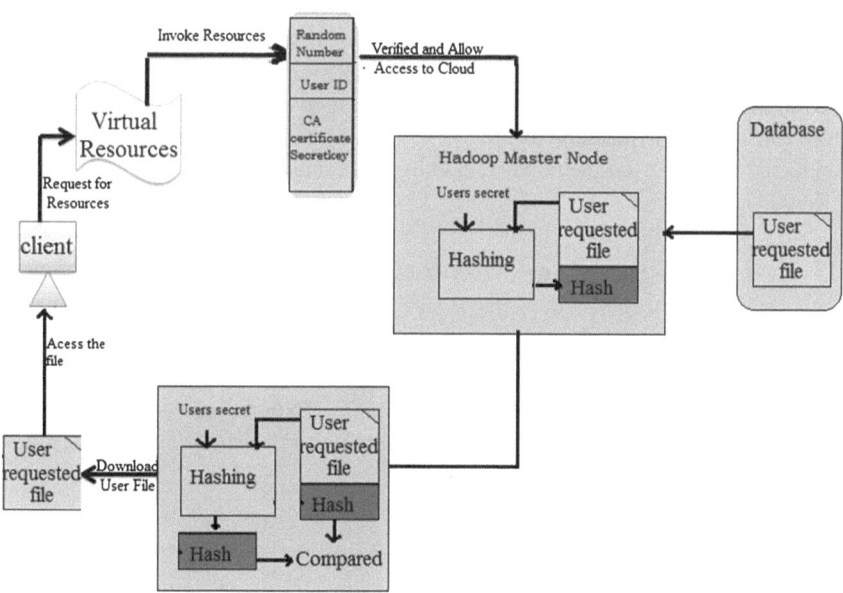

Fig. 3 Overall architecture of hashing-based cloud environment

Fig. 4 Average running time and collision rate for various hash algorithms

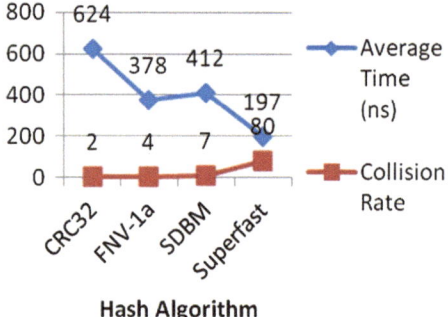

3.4 Non-cryptographic Model

For some applications, the integrity and authenticity of the data is the only concern and data confidentiality is not a problem. This can be achieved by hashing. As mentioned earlier, whenever a user is registered into the Hadoop database, the user will be assigned with a secret key and will be converted into hash and given to Hadoop client node. Whenever the user requested a file, the Hadoop master node will generate hash of the requested file and the respected user's secret key, appended to the file and sent to the Hadoop client node.

The Hadoop client node will check the received data, and it will generate its own hash using the data portion of the received file and user's secret and will compare it with received hash, and if both are same, then the file will be passed to user. Figure 4 shows the average time needed and collision rate for various algorithms; even though superfast algorithm is taking less time in execution, its collision rate is more. Hence, FNV hashing technique is considered and the experimentation is conducted by using the parameters: running time and collisions.

3.5 Hadoop Cluster Load Monitoring

The size and the load of Hadoop cluster can be very dynamic. Depending upon the jobs that Hadoop is running, the load of the nodes can be changed. For monitoring of Hadoop cluster, in this experimentation, Ganglia, an open source software tool, is used, which offers a complete solution for monitoring, visualizing, and archiving the system metrics.

4 Evaluation Model

To measure the monitoring system, there are many ways, among which we used two models to evaluate ECMF, which are accuracy and consumption. We discuss accuracy model in Sect. 4.1 and consumption model in Sect. 4.2.

4.1 Accuracy

The accuracy of the monitoring information gives the error between the collected and the actual data.

Assume there are N metrics available at system that we want to monitor. X_1, X_2, ... X_N are the maximum measure of the N metrics. V_1, V_2, ... V_N are the collected measure of the N metrics. A_1, A_2, ... A_N are the actual measure of the N metrics.

Then, the accuracy of the monitoring information is

$$A = 1 - \frac{1}{N} \sum_{i=1}^{N} \frac{[V_i - A_i]}{X_i}$$

The metrics are memory, CPU and disk usage, and workload on the network.

4.2 Consumption

The task of monitoring middleware will consume resources. The overall usage can be given as the total cost of CPU, memory, disk usages, and workload on network, represented as (u_1), (u_2), (u_3), and (u_4), respectively.

$$T_{\text{cons}} = \sum_{i=4}^{4} U_i * P_i$$

where P_i is the cost of the Amazon EC2 pricing that has been referenced from.

5 Experimental Results

5.1 Environment Setup

To verify performance of proposed system, a Hadoop cluster of one name node and three data nodes is selected. They are configured with 2.6 GHz processor and 4 GB/ 500 GB and used high-speed switches to connect all these. The OS used is Linux 4.1, and version of Hadoop used is 0.20.1, and for GNU, Java 1.6.0 is installed in all nodes. For virtualization and establishing infrastructure, we used Xen hypervisor and eucalyptus. And for performance monitoring, Ganglia-3.6.0 is used.

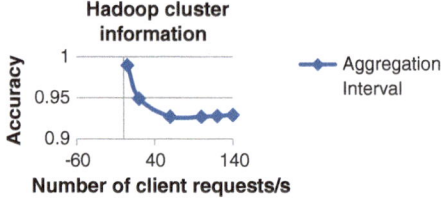

Fig. 5 Accuracy changes with number of nodes and aggregation level

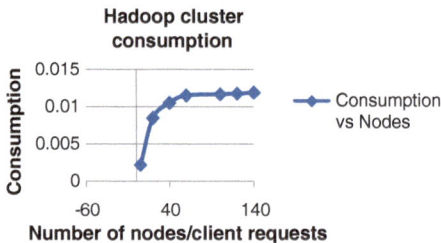

Fig. 6 Consumption level is decreasing as client requests are increasing

5.2 Experiment to Test Accuracy

In this experiment, we considered only the important metrics: CPU, memory, and disk usage and network workload. Figure 5 shows the metric accuracy, which varies with aggregation interval and number of nodes in the cluster. With the increase in aggregation interval, the accuracy of monitoring information will decrease, but at a certain point, direction is changed toward increase in accuracy. The accuracy is also increasing as the number of nodes in the cluster is increasing.

5.3 Consumption Experiments

Figure 6 shows the consumption with different number of nodes or client requests. As the number of nodes is very less, consumption is high; it gradually reached stable state and then reduced gradually as the node number is increasing.

6 Conclusion and Future Work

In this paper, we presented a new methodology of providing integrity and authenticity for the data outsourced by the user over cloud using FNV, a non-cryptographic hashing technique. This paper implemented an efficient and secure method for attribute-based encryption. We introduced two techniques: encryption of log file and message digest that can only be read by those who have credentials. We also represented the results of processing time for different algorithms with varying collisions. Finally, this paper gave a model to evaluate cluster. The model evaluates the cloud system in terms of resource consumption and accuracy of the result.

We want to develop the evaluation model in future with the extension of secure ranked keyword search technique to identify the data of user interest in cloud environment.

References

1. http://www.edureka.in/blog/what-is-big-data-and-why-learn-hadoop/
2. http://tools.ietf.org/html/draft-eastlake-fnv-07
3. Svantesson D, Clarke R (2010) Privacy and consumer risks in cloud computing. Comput Law Secur Review 26(4):391–397
4. King NJ, Raja VT (2012) Protecting the privacy and security of sensitive customer data in the cloud. Comput Law Secur Rev 28(3):308–319
5. Breitinger F, Stivaktakis G, Baier H (2013) A framework to test algorithms of similarity hashing. Digit Invest 10:S50–S58
6. Rupesh M, Chitre DK (2012) Data leakage and detection of guilty agent. Int J Sci Eng Res 3 (6)
7. Hadoop, http://hadoop.apache.org
8. http://www.isthe.com/chongo/tech/comp/fnv/index.html#history
9. Zhao J, Wang L, Tao J, Chen J, Sun W, Ranjan R, Kołodziej J, Streit A, Georgakopoulos D (2014) A security framework in GHadoop for bigdata computing across distributed Cloud data centers. Comput Syst Sci 80:994–1007
10. Wang L, Tao J, Ranjan R, Marten H, Streit A, Chen D, Chen J (2013) G-Hadoop: mapreduce across distributed data centers from data-intensive computing. Future Gener Comput Syst 29 (3):739
11. Caballer M, de Alfonso C, Molto G, Romero E, Blanquer I, Garcia A (2014) Code cloud: A platform to enable execution of programming models on the Clouds. J Syst Softw 93:187–198
12. AL-Saiyd NA, Sail N (2013) Data integrity in cloud computing security. Theor Appl Inform Technol 58
13. Dillibabu M, Kumari S, Saranya T, Preethi R (2013) Assured protection and veracity for cloud data using Merkle hash tree algorithm. Indian J Appl Res 3:1–3
14. Mounika CH, RamaDevi L, Nikhila P (2013) Sample load rebalancing for distributed hash table in cloud. ISRO J Comput Eng 13:60–65